THE MAJOR TRANSITIONS IN E

D1423824

THE
MAJOR TRANSITIONS IN
EVOLUTION

■

John Maynard Smith and Eörs Szathmáry

Oxford New York Tokyo

OXFORD UNIVERSITY PRESS

Oxford University Press, Great Clarendon Street, Oxford OX2 6DP

Oxford New York

Athens Auckland Bangkok Bogota Bombay
Buenos Aires Calcutta Cape Town Dar es Salaam
Delhi Florence Hong Kong Istanbul Karachi
Kuala Lumpur Madras Madrid Melbourne
Mexico City Nairobi Paris Singapore
Taipei Tokyo Toronto Warsaw

and associated companies in
Berlin Ibadan

Oxford is a trade mark of Oxford University Press

Published in the United States by
Oxford University Press Inc., New York

First published by W. H. Freeman/Spektrum, Oxford 1995
and reprinted (with corrections) 1995 (twice)
First published by Oxford University Press 1997

A catalogue record for this book is available from the British Library

Library of Congress Cataloging in Publication Data
(Data available)
ISBN 0 19 850294 X

Printed in Great Britain by
Bookcraft (Bath) Ltd
Midsomer Norton, Avon

To our parents

Contents

Tables

Preface

THIS book is about the origin of life, of the genetic code, of cells, of sex, of multicellular organisms, of societies, and of language. Such a book is inevitably speculative, because it is an account of a series of unique events that happened a long time ago. But these are matters on which we must speculate. Why else would we study evolution? It is true that evolutionary biology has some practical relevance—for example, to animal breeding or to the origins of antibiotic resistance—but the real reason why we study it is that we are interested in origins. We want to know where we came from.

Although the book is speculative, however, we think that it is a contribution to science, and not to fantasy. Speculation is constrained in two ways. First, each event must be explained in a way that is consistent with a general theory of evolutionary change, the theory of evolution by natural selection. Second, an adequate account of the origin of any system must explain the peculiarities of that system as it exists today: for example, a theory of the origin of the code should explain why it is a triplet code, why it is redundant, why similar codons specify chemically similar amino acids, and so on. In other words, theories about origins can be tested by looking at the present.

This means that we have had to think very hard about the essential characteristics of the various levels of organization whose origins we are trying to explain. We have had to learn a lot of biology, from molecular genetics to linguistics, to write the book. We think that anyone who reads it will also learn some biology.

We have to thank Barry Cox for catalysing our first encounter at the ICSEB III in 1985 at Brighton. The basic ideas behind the book first emerged in 1987–8, when E.S. spent some time with J.M.S. at the University of Sussex. The plan actually to write a book, however, was not conceived until 1991, when E.S. was at the National Institute for Medical Research in London. It seemed that we might be able to collaborate fruitfully because we have a common conviction that heredity and selection are central to evolutionary explanations, but otherwise tend to approach biological problems from different ends— respectively, from chemistry and the physical sciences, and from natural history. Most of the time we have been working in different countries. We have shared the job of writing preliminary drafts of the various chapters, but the final version of every chapter is very much a joint effort, which we have argued about on the many occasions when we have met. There remain a few disagreements between us about what should have been included—E.S. always wanted more

chemistry than J.M.S. could understand — but no significant disagreement about ideas.

While the book was being written, E.S. was a guest at various institutions abroad. Tom Kirkwood, then head of the Laboratory of Mathematical Biology at the National Institute for Medical Research, was very tolerant of this not very MRC-like activity, and he was also a keen discussant. The Institute for Advanced Study in Berlin (Wissenschaftskolleg zu Berlin) provided E.S. with luxurious working and financial conditions for 10 months, where he could discuss many topics with Peter Hammerstein and James Griesemer. Finally, a guest professorship at the Institute of Zoology, University of Zürich, gave the opportunity to teach a course on theoretical biology, largely based on the manuscript of this book. Rüdiger Wehner, as head of the institute, supported this strongly.

The colleagues with whom we have discussed particular problems are too numerous to mention individually, but a few must be named. Before starting on the chapters dealing with development, we had a splendid two-day tutorial with Jonathon Cooke. An earlier draft, excepting the last two chapters, was read by Laurence Hurst, Stephen Kearsey and Mark Ridley. Their comments were invaluable, not only in pointing out errors but in suggesting ways in which things could be explained more clearly. We have not always taken their advice: it is impossible to resist mentioning one example when we did not. We drew a parallel between Eigen's notion of an error threshold and a phase transition. One of the three wrote 'This does NOT remind me of a phase transition: it cannot be too widely known that nothing reminds me of a phase transition'. Believing, as we do, that the only thing required to make this particular commentator the complete evolutionary biologist is a love of phase transitions, we have left the remark as it stood.

Michael Rodgers has been unfailingly helpful, not least in finding three such wise and knowledgeable referees. He was also responsible for finding Sarah Bunney, who copyedited the manuscript with great care, and on occasions made helpful suggestions on matters of substance, and Jane Templeman, who prepared the figures from our often rather scruffy sketches.

Our greatest debt, however, is to the two men, J. B. S. Haldane and Tibor Gánti, who taught us. Some of their ideas are referred to in the text, but their influence is present on every page. Without their teaching, we might never have tried to write the book at all.

1 Introduction

1.1 Preamble

Living organisms are highly complex, and are composed of parts that function to ensure the survival and reproduction of the whole. This book is about how and why this complexity has increased in the course of evolution. The increase has been neither universal nor inevitable. Bacteria, for example, are probably no more complex today than their ancestors 2000 million years ago. The most that we can say is that some lineages have become more complex in the course of time. Complexity is hard to define or to measure, but there is surely some sense in which elephants and oak trees are more complex than bacteria, and bacteria than the first replicating molecules.

Our thesis is that the increase has depended on a small number of major transitions in the way in which genetic information is transmitted between generations. Some of these transitions were unique: for example, the origin of the eukaryotes from the prokaryotes, of meiotic sex, and of the genetic code itself. Other transitions, such as the origin of multicellularity, and of animal societies, have occurred several times independently. There is no reason to regard the unique transitions as the inevitable result of some general law: one can imagine that life might have got stuck at the prokaryote or at the protist stage of evolution.

There are obvious difficulties in discussing unique events that happened a long time ago. How can we ever know that our suggested explanations are correct? After all, historians cannot agree about the causes of the Second World War. We accept that certainty is impossible, but there are several reasons why we think the enterprise is worth while. First, we have one great advantage over historians: we have agreed theories both of chemistry and of the mechanism of evolutionary change. We can therefore insist that our explanations be plausible both chemically, and in terms of natural selection. This places a severe constraint on possible theories. Indeed, the difficulty often lies, not in choosing between rival theories, but in finding any theory that is chemically and selectively plausible. Further, theories are often testable by looking at existing organisms.

A second reason why the study of origins is worth while is that, to understand the origin of some structure, one must first understand what is essential about it — what features it must have if it is to work at all. Writing this book has forced us to learn a lot of biology, from the nature of the genetic code to the nature of human language. But the major reason for thinking about origins is more intangible: we want to know the answers.

In this introduction, we give an outline of the rest of the book. But, first, we discuss two preliminary topics. In section 1.2, we explain why we do not regard the evolutionary process as one of inevitable progress; in section 1.3, we ask how complexity might be defined and measured. Unfortunately, no very illuminating answers to these questions are possible. Readers who already know that evo-

lution is not equivalent to progress, and that complexity is hard to measure, can safely skip to the start of section 1.4.

In section 1.4, we list the major transitions. The justification for discussing such an apparently diverse series of changes in a single book is that they have features in common. The most important of these is that entities that were capable of independent replication before the transition can replicate only as part of a larger whole after it. This common feature raises a common problem: why does not selection between entities at the lower level disrupt integration at the higher one? These points are explained in more detail in section 1.5. Other features that are common to different transitions are described in section 1.6.

1.2 The fallacy of progress

The notion of progress has a bad name among evolutionary biologists. Lamarck accepted the earlier idea of a ladder of nature, and argued that organisms have an inherent tendency to climb the ladder. It was Lamarck's notion of an inherent tendency, rather than his belief in the inheritance of acquired characters, that Darwin was rejecting when he said that his theory had nothing in common with Lamarck's: he rightly saw that to explain evolution by an inherent tendency is as vacuous as to say that a man is fat because he has an inherent tendency to obesity. Today, we are unhappy with a picture of evolution that places us at the summit, and arranges all other organisms in a line behind us: what have we to be so proud about? To be fair, humans were by no means at the summit of the medieval *scala naturae*; there were angels and archangels above us as well as worms below.

There are, of course, more solid reasons, both empirical and theoretical, for rejecting a simple image of progress on a linear scale. Empirically, the history of life is better visualized as a branching tree than as a single ascending line. The fossil record shows that many organisms — horseshoe crabs, the coelacanth, crocodiles, for example — have undergone little change, progressive or other-wise, for hundreds of millions of years. On a shorter timescale, sibling species tell the same story. The fruit flies *Drosophila melanogaster* and *D. simulans* are hard to distinguish morphologically, but molecular data indicate that they are separated by several million years of evolution. Hence, either morphological evolution in the two species has been almost exactly parallel, which is implausible, or neither species has changed.

On the theoretical side, there is no reason why evolution by natural selection should lead to an increase in complexity, if that is what we mean by progress. At most, the theory suggests that organisms should get better, or at least no worse, at doing what they are doing right now. But an increase in immediate 'fitness' — that is, expected number of offspring — may be achieved by losing eyes or legs as well as by gaining them. Even if an increase in fitness cannot be equated with

an increase in complexity, or with progress, it might seem at first sight that R. A. Fisher's (1930) 'fundamental theorem of natural selection' at least guarantees an increase in fitness. The theorem states that the rate of increase in the mean fitness of a population is equal to the genetic variance in fitness: since variances cannot be negative, the theorem states that fitness can only increase. If so, 'mean fitness' in biology is an analogue of entropy in physics: it gives an arrow to time. Thus in physics the inevitable increase in entropy distinguishes past from future: if mean fitness can only increase, this gives a direction to evolution. It seems that Fisher did indeed think that his theorem could play such a role: otherwise, why 'fundamental'? Unfortunately, the theorem holds only if the relative fitnesses of genotypes are constant, and independent of their frequencies in the population: for many traits, such constancy does not hold.

1.3 The measurement of complexity

Even if progress is not a universal law of evolution, common sense does suggest that at least some lineages have become more complex. How might one measure this increase? A possible answer is in terms of the DNA content of the genome, which can be thought of as instructions for making the organism: more complex organisms require lengthier instructions. If we look at total DNA, we reach the rather demeaning conclusion that a lungfish or a lily is some 40 times as complex as a human (Table 1.1). Things make more sense if we allow for the fact that a varyingly small proportion of the DNA codes for anything. It then appears that eukaryotes have more coding DNA than prokaryotes (although the difference between yeast and *Escherichia coli* is small); that higher plants and invertebrate animals have more DNA than single-celled organisms; that an invertebrate with wings, legs and eyes has more DNA than a nematode; and that

Table 1.1 Genome size and DNA content

	Genome size (base pairs \times 10^{-9})	Coding DNA (%)
Bacterium (*Escherichia coli*)	0.004	100
Yeast (*Saccharomyces*)	0.009	70
Nematode (*Caenorhabditis*)	0.09	25
Fruit fly (*Drosophila*)	0.18	33
Newt (*Triturus*)	19.0	1.5–4.5
Human	3.5	9–27
Lungfish (*Protopterus*)	140.0	0.4–1.2
Flowering plant (*Arabidopsis*)	0.2	31
Flowering plant (*Fritillaria*)	130.0	0.02

Data from Cavalier-Smith, 1985.

vertebrates have more DNA than invertebrates. This last observation is rather puzzling: it is possible that the large brains of vertebrates require large genomes.

These data make sense, but what they tell us about structural and functional complexity is very limited. At present, we can say rather little about how much genetic information is needed to program a particular morphological structure.

1.4 The major transitions

A tentative list of the major stages in the evolution of complexity, and the transitions between them, is given in Table 1.2. We have confined our attention to the way in which information is transmitted between generations, so have not included major phenotypic changes, such as the conquest of land by plants and animals, or the origins of vision, or of flight or of homiothermy, which did not involve such a change in the method of information transmission.

One feature is common to many of the transitions: entities that were capable of independent replication before the transition can replicate only as part of a larger whole after it. Some examples will make this clearer:

The origin of chromosomes. Initially, there were independently replicating nucleic acid molecules: after the transition, a set of linked molecules must replicate together.

The origin of eukaryotes. The ancestors of mitochondria and chloroplasts were once free-living prokaryotes: today, they can replicate only within a host cell.

The origin of sex. The first eukaryotes could, presumably, reproduce asexually, and on their own: today, most eukaryotes can replicate only as part of a sexual population.

The origin of multicellular organisms. The cells of animals, plants and fungi are descended from single-celled protists, each of which could survive on its own: today, they exist (outside the laboratory) only as parts of larger organisms.

Table 1.2 The major transitions

Replicating molecules	→	Populations of molecules in compartments
Independent replicators	→	Chromosomes
RNA as gene and enzyme	→	DNA + protein (genetic code)
Prokaryotes	→	Eukaryotes
Asexual clones	→	Sexual populations
Protists	→	Animals, plants, fungi (cell differentiation)
Solitary individuals	→	Colonies (non-reproductive castes)
Primate societies	→	Human societies (language)

The origin of social groups. Individual ants, bees, wasps and termites can survive and transmit genes (either their own or ones genetically similar to their own) only as part of a social group: the same is effectively true of humans.

It might be asked why we do not include the origin of ecosystems in our list of transitions. There are two reasons. First, in temporal terms an ecosystem is not the final stage in a series: ecosystems are as old as replicating molecules. Second, ecosystems are not individuals, separated from others, whereas the other stages we have listed (including sexual species, and insect colonies) do have a degree of individuality, and separateness from other entities of the same kind. For this reason, ecosystems cannot be units of selection.

Given this common feature of the major transitions, there is a common question we can ask of them. Why did not natural selection, acting on entities at the lower level (replicating molecules, free-living prokaryotes, asexual protists, single cells, individual organisms), disrupt integration at the higher level (chromosomes, eukaryotic cells, sexual species, multicellular organisms, societies)? It is because there is this common question that we have found it illuminating to compare the different transitions. In fact, one of the stimuli for attempting the work was our realization that a model one of us had developed to analyse the origin of compartments containing populations of molecules was formally and mathematically similar to a model that the other had developed to analyse the evolution of cooperative behaviour in higher animals.

First, we must establish that the problem is not an imaginary one: that there is a real danger that selection at the lower level will disrupt integration at the higher. This is best done by listing examples (Table 1.3) in which such a process can be observed today:

• If Mendel's laws are rigorously obeyed, a gene can only increase its representation in future generations by ensuring the success of the cell in which it finds itself, and of the other genes in the cell. Hence, Mendel's laws ensure the evolution of cooperative, or 'coadapted', genes. But the laws are broken, in meiotic drive, and by transposable elements. These are examples of the more general phenomenon of intragenomic conflict, which is the topic of Chapter 10.

Table 1.3 Conflict between selection at different levels

Form of cooperation	Exceptions
A fair meiosis	Meiotic drive, transposition
Sexual reproduction	Parthenogenesis
Differentiation of somatic cells	Escape from growth control
Non-reproductive castes of social insects	Egg-laying worker bees

- A sexual population has an advantage, in rate of evolution and in the elimination of harmful mutations, over an asexual one. But a parthenogenetic female has, in the short run, a twofold advantage over a sexual one, and parthenogens are not uncommon.

- A gene in a somatic cell of a plant might best ensure the transmission of replicas of itself by giving rise to a flower bud, even if this reduced the success of the whole plant.

- A bee colony produces more reproductives if the workers raise the queen's offspring. But workers do lay eggs (which are unfertilized, and hence male).

We cannot hope to explain these transitions in terms of the ultimate benefits they conferred. For example, it may be that, in the long run, the most important difference between prokaryotes and eukaryotes is that the latter evolved a mechanism for chromosome segregation at cell division that permits DNA replication to start simultaneously at many origins, whereas prokaryotes have only a single origin of replication. At the very least, this was a necessary precondition for the subsequent increase in DNA content, without which complexity could not increase. But this is not the reason why the change occurred in the first place: as we explain in Chapter 6, the new segregation mechanism was forced on the early eukaryotes by the loss of a rigid cell wall, which plays a crucial role in the segregation of prokaryotic chromosomes. Or to take a second example, meiotic sex was an important preadaptation for the subsequent evolutionary radiation of the eukaryotes, but it could not have originated for that reason.

The transitions must be explained in terms of immediate selective advantage to individual replicators: we are committed to the gene-centred approach outlined by Williams (1966), and made still more explicit by Dawkins (1976). There is, in fact, one feature of the transitions listed in Table 1.2 that leads to this conclusion. At some point in the life cycle, there is only one copy, or very few copies, of the genetic material: consequently, there is a high degree of genetic relatedness between the units that combine in the higher organism. The importance of this general principle was first emphasized by Hamilton (1964) in his explanation of the evolution of social behaviour, but we believe it to be quite general. To give two other examples: multicellular organisms develop from a single fertilized egg, so that their cells are genetically identical, except for somatic mutation; most eukaryotes inherit their organelles from one parent only, so that the organelles in a single individual are almost always genetically identical. We think that a similar principle operated in the origin of the earliest cells.

The principle of genetic similarity resulting from a small number of founders is important at the time of the transition. Two other processes — contingent irreversibility and central control — help to explain the maintenance of higher-level entities, once they have arisen, although they are less relevant to the origin of such entities.

Contingent irreversibility

If an entity has replicated as part of a larger whole for a long time, it may have lost the capacity for independent replication that it once had, for accidental reasons that have little to do with the selective forces that led to the evolution of the higher-level entity in the first place. For example, mitochondria cannot resume independent existence, if only because most of their genes have been transferred to the nucleus; cancer cells may escape growth control, but have no independent future as protists; worker bees may lay male eggs, but cannot establish a new colony on their own.

The contingent nature of irreversibility is perhaps best illustrated by the reversion from sex to parthenogenesis. Mammals are never parthenogens, probably because, at some loci in some tissues, only the allele inherited from the father is active: hence, in an embryo with no father, some essential gene activities are missing. Gymnosperms are also never parthenogens, perhaps for a different reason: chloroplasts are transmitted in the pollen. Anamniote vertebrates (fish and amphibians), although they may be parthenogens, always require sperm from the males of another species to initiate development, perhaps because the sperm provides a centriole. The relevance of these sexual hang-ups — and there are many others — is to show how various and accidental are the reasons why reversal is difficult or impossible.

Three points must be made about irreversibility:

- Different mechanisms may be involved in the origin and the maintenance of higher-level organization. It would be absurd to suggest sexual imprinting as a cause of the origin of sex, or gene transfer to the nucleus as a cause of the symbiotic origin of mitochondria.

- Irreversibility is not absolute. There are, after all, many successful ameiotic parthenogens. To give a second example, viruses and transposons are probably the descendents of well-behaved chromosomal genes. Irreversibility, therefore, is not only irrelevant to the origin of higher-level organization: it is not a sufficient explanation for its maintenance.

- The fact that a once independent replicator can no longer revert does not mean that selfish mutations cannot occur in the genes of such replicators.

Central control

Analogy with human society might suggest that the maintenance of organization depends on some kind of central control: people pay taxes because they would be punished if they did not. But the analogy is not a close one. In human societies, central control depends either on the existence of an armed group within society — for example, a feudal aristocracy — or on a consensus imposed by a majority, or on some combination of the two. The notion of armed force has little application in biology — there are not armed and unarmed genes. Although

in primitive social insects the reproductive queen may acquire and maintain her position by force, this is not so in more highly evolved species. Nor is the idea of a majority consensus particularly helpful: genes do not vote. But there is one sense in which the idea of central control may be helpful. If a 'selfish' mutation occurs in a chromosomal gene, a suppressor mutation at any other locus in the genome would be favoured by selection. Hence, the rest of the genome may win the contest, not because of any analogue of majority voting, but because of the large number of loci, and hence of possible suppressor mutations, that are available for each selfish mutation. It may be relevant that attempts to use driving chromosomes in biological control have so far failed because of the rapid evolution of suppression. It is in this sense that Leigh's (1971) idea of a 'parliament of genes' should be taken.

1.5 Duplication, symbiosis and epigenesis

In Figs 1.1–1.3, we illustrate three ways in which genetic complexity may increase:

Duplication and divergence. After the origin of the eukaryotes, subsequent increase in genetic information has probably occurred in this way. The mere duplication of a gene adds no new information, but the divergence of the two copies does so. Gene families, such as the haemoglobin family, illustrate this process. Important as it has been in the later evolution of the eukaryotes, we do not think it played a significant role in the major transitions themselves.

Symbiosis. The significance of symbiosis in the origin of the eukaryotes (Margulis, 1970) is now familiar. We think it was also the mechanism whereby chromosomes first evolved: that is, chromosomes arose by the linking together of genes that were different at the outset, and not by duplication and divergence. The significance of symbiosis in some other evolutionary innovations is discussed in Chapter 11. There are, however, important transitions that did not arise in this way. The differentiated cells of a multicellular organism are not descended from different ancestors: nor are the castes of social insects.

Epigenesis. Genetic change, in the sense of altered DNA sequence, is not required for the differentiation of somatic cells, or of the castes of social insects. Instead, different genes are active in different cells, or in different individuals.

Fig. 1.1 Increase in complexity by duplication, followed by divergence.

Fig. 1.2 Increase in complexity by symbiosis, followed by compartmentation, and synchronized replication.

Fig. 1.3 Increase in complexity by epigenesis.

different genes are active in different cells

1.6 Some other features of the major transitions

The idea of levels of organization, and hence of levels of selection, is central to this book. Not all major transitions, however, can be analysed in this way. Perhaps the most important transition of all is that between organisms in which both genetic material and enzymes were made of RNA (the RNA world) and modern organisms in which the genetic material is DNA and enzymes are pro- teins — a division of labour that requires that there be coding and translation. A second transition, which also involves a change in the language whereby in- formation is transmitted and in the physical medium that carries that language, is the origin of human speech. We accept this as being the decisive step in the origin of specifically human society.

We will meet these two characteristics — the division of labour and a change in language — repeatedly.

The division of labour

To survive, organisms must complete different tasks. Every organism has many components — molecules, cells, segments, organs, and so on. In the evolu- tionary history of these components, an initially identical set of objects often becomes differentiated and functionally specialized. A division of labour can also occur between organisms within a population. Why should selection favour such a division?

The underlying principle can be illustrated by the evolution of dioecy from hermaphroditism (for details, see Maynard Smith, 1978). There are obvious advantages associated with hermaphroditism, so why are separate sexes so common? Let R_m and R_f be the reproductive success of males and females, respectively, and the reproductive success of a hermaphrodite be αR_m as a male and βR_f as a female. It is easy to show that dioecy will be favoured if $\alpha + \beta < 1$. The main reason why this should be so is the existence of specialized organs, useful only to a male or to a female. Thus imagine a hermaphrodite red deer, investing half as much as a typical male in weapons and excess size, and half as much as a typical female in nourishing the young. In all probability, such a hermaphrodite would father less than half as many offspring as a typical male ($\alpha < 1/2$), and would mother less than half as many as a typical female ($\beta < 1/2$): that is, $\alpha + \beta < 1$, and hermaphroditism would not pay. The stability of dioecy depends on the efficiency of specialized organs.

Molnár (1993, 1994) has noted some other examples, which have a similar explanation. Our list is as follows:

• The appearance of many specific enzymes from a set of multifunctional low- efficiency enzymes.

- In the RNA world, RNA serves as both genetic material and catalyst: in our world, DNA is the genetic material, and most enzymes are proteins.

- In prokaryotes, there is a single cell compartment, whereas in eukaryotes the genetic nucleus and metabolic cytoplasm are separated, and additional organelles evolved, some recruited symbiotically.

- In sexual populations, isogamy has repeatedly evolved to anisogamy, with differentiated sperm and ova.

- Differentiation between genetically transmitted germ and mortal soma has arisen repeatedly: nuclear differentiation in ciliate protozoans has a similar function to germ-line segregation in animals.

- Hermaphrodites are replaced by separate sexes.

- In eusocial insects, castes appeared, some of which are non-reproductive. There is a clear analogy between a non-reproductive caste and organismal soma. In some cases, there is differentiation between non-reproductives.

New ways of transmitting information

Heredity means that like begets like: it requires some means whereby information can be transmitted. A crucial distinction (section 4.2) is between systems of 'limited heredity', in which only a few distinct states can be transmitted, and systems of 'unlimited heredity', capable of transmitting an indefinitely large number of messages. We discuss the following events:

- The origin of simple autocatalytic systems (networks, short oligonucleotide analogues) with limited heredity (sections 3.4, 3.6, 4.2 and 5.3).

- The origin of template replication of polynucleotide-like molecules, providing unlimited heredity (sections 5.4 and 5.5).

- The origin of the genetic code in the context of the RNA world, prior to translation. Coding serves only to equip amino acids with unambiguous trinucleotide handles (section 6.3).

- The origin of translation and encoded protein synthesis (section 6.3).

- The emergence of hereditary regulative states in prokaryotes and simple eukaryotes (section 13.2).

- The evolution of epigenetic inheritance with unlimited heredity: the emergence of animals, plants and fungi (section 13.2).

- The emergence of protolanguage in *Homo erectus* — a cultural inheritance system with limited potential in which only certain types of statement can be made.

- The emergence of human language with a universal grammar and unlimited semantic representation.

Certainly we do not answer all the questions we ask, but sometimes it is as useful to ask a new question as to answer an old one.

2 What is Life?

2.1 The definition of life

Imagine that, when the first spacemen step out of their craft onto the surface of one of the moons of Jupiter, they are confronted by an object the size of a horse, rolling towards them on wheels, and bearing on its back a concave disc pointing towards the Sun. They will at once conclude that the object is alive, or has been made by something alive. If all they find is a purple smear on the surface of the rocks, they will have to work harder to decide. This is the phenotypic approach to the definition of life: a thing is alive if it has parts, or 'organs', which perform functions. William Paley explained the machine-like nature of life by the existence of a creator: today, we would invoke natural selection.

There are, however, dangers in assuming that any entity with the properties of a self-regulating machine is alive, or an artefact. In section 2.2, we tell the story of a self-regulating atomic reactor, the Oklo reactor, which is neither. This story can be taken in one of three ways. First, it shows the dangers of the phenotypic definition of life: not all complex entities are alive. Second, it illustrates how the accidents of history can give rise spontaneously to surprisingly complex machine-like entities. The relevance of this to the origin of life is obvious. In essence, the problem is the following. How could chemical and physical processes give rise, without natural selection, to entities capable of hereditary replication, which would therefore, from then on, evolve by natural selection? The Oklo reactor is an example of what can happen. Finally, section 2.2 can simply be skipped: the events were interesting, but do not resemble in detail those that led to the origin of life on Earth.

There is an alternative to the phenotypic definition of life. It is to define as alive any entities that have the properties of multiplication, variation and heredity. The logic behind this definition, first proposed by Muller (1966), is that a population of entities with these properties will evolve by natural selection, and hence can be expected to acquire the complex adaptations for survival and reproduction that are characteristic of living things. Such a definition, however, is inadequate if we want to understand the origin of life. One way of seeing this is to consider the experiments of Spiegleman (1970) and others on the evolution of RNA molecules in test tubes. In these experiments, which are described in greater detail in Chapter 4, a test tube containing the four ribonucleotides, and an enzyme, $Q\beta$ replicase, is seeded with molecules of RNA. These molecules are replicated, with errors, and hence the population evolves. By serially transferring a sample of the population to a new tube, molecules can be obtained that are adapted to replicate in particular environments—for example, in the presence of inhibitory drugs. Now, by Muller's definition, these molecules are alive, but they do not represent a plausible early stage in the origin of life: one fact about the primitive Earth of which we are reasonably confident is that there were no programmed protein replicases. To explain the origin of life requires that we explain the origin of metabolism as well as of replication.

We are therefore faced with two alternative definitions of life, phenotypic and hereditary. The essential phenotypic feature of the first living things was metabolism—linked series of chemical reactions, driven by an extrinsic source of energy. In all existing organisms, heredity depends on the template replication of polymers. It seems to us likely that this has been the case from the beginning. What were the connections between the two processes, metabolism and heredity? In existing organisms, there are two such connections:

- metabolism supplies the monomers from which the replicators are made, and
- replicators alter the kinds of chemical reactions occurring in metabolism: only if this is true can natural selection, acting on replicators, influence the evolution of metabolism.

How could such a two-way interaction arise? The simplest assumption is that, initially, metabolism consisted only of abiotic chemical reactions, of the kind studied by Miller and Urey, as described in the next chapter. Such an abiotic metabolism supplied the monomers from which the first replicating polymers were synthesized, but those replicators did not themselves influence metabolism. Natural selection would act solely to select those replicators that multiplied best, in a chemical environment that they could not alter. This is a 'replicator-first', or 'genes-first', model of the origin of life. Only later did replicators evolve the ability to alter their own chemical environment.

An alternative scenario is to dispense with specific replicating molecules, and to suppose that the metabolic system itself acquired a kind of heredity. We do not regard this alternative as particularly plausible, but we discuss it in some detail in Chapter 5 because it is the main alternative to the view we favour.

In section 2.3, we describe the simplest system yet proposed, the chemoton, which includes both a metabolic system and a replicating template, and a two-way interaction between them.

2.2 The Oklo reactor

The first artificial atomic fission reactor, created by Enrico Fermi and his colleagues, started to work in 1942 in Chicago. One isotope of uranium, ^{235}U, is used to fuel such reactors. By absorbing a neutron, the nucleus of a ^{235}U atom is raised to an excited and unstable state. The atom relieves itself from this instability by splitting into two daughter nuclei, at the same time emitting high-energy neutrons and gamma rays. A chain reaction starts when at least one of the emitted neutrons is captured by another ^{235}U atom, which in turn decays and emits neutrons, and so on. The process generates heat, which can be used to warm water and make electricity.

France has been obtaining uranium ore from Gabon in West Africa. In 1972, a load of ore carried only 0.7171 per cent of ^{235}U, instead of the usual 0.7202

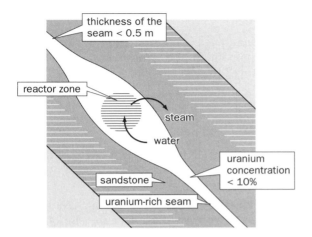

thickness of the
seam < 0.5 m

reactor zone

steam

water

sandstone

uranium-rich seam

uranium
concentration
< 10%

Fig. 2.1 A natural nuclear reactor. At Oklo, there is a uranium-rich seam, within a layer of sandstone. The process whereby a controlled reaction occurred is explained in the text.

per cent. Sabotage and stealing were immediately suspected, but not confirmed. It turned out that a natural fission reactor had been operating. This reactor has been called Oklo: how could it have arisen?

The first stage (Fig. 2.1) was the wide dispersal of chemically reduced uranium in plutonic rocks. Next, as a result of weathering, uranium accumulated in river beds, still in a relatively reduced form. Roughly 2000 million years ago, the concentration of oxygen in the air began to rise rapidly as a result of photosynthesis by cyanobacteria. In consequence, much of the uranium was oxidized. Since oxidized uranium is more soluble, it was carried to the delta of an ancient river. Deltas tend to be reducing, because of the rotting of organic sediments. Hence, uranium was precipitated, and covered by later sediments. The river has since disappeared, but the delta remains.

Merely to concentrate the uranium was not enough to produce a reactor. At Oklo, uranium ore embedded in sandstone was lying on a layer of granite. This granite layer was then tilted to roughly 45°, together with the overlying sandstone. Clefts formed in the sandstone, and percolating (oxidizing) water further concentrated the uranium in certain places. Once enough uranium had accumulated, the chain reaction could start. For this to happen, uranium must have reached a concentration above 10 per cent. The ^{235}U layer could not have been too thin, otherwise too many neutrons would have escaped. Further, elements that act as poisons by absorbing neutrons had to be absent, as in fact they are in the relevant layers. Some moderator material was needed to slow down neutrons so that they were not absorbed by ^{238}U: this moderating material was water itself. Perhaps most surprising of all, the reaction had to be self-regulating. This regulation was also performed by water. If the reactor ran too fast, water was turned into steam, and the reactor slowed down in the absence of the moderating effect of the water: as the reactor slowed, water condensed, and prevented the absorption of neutrons by ^{238}U, thus speeding up the reaction.

Two general conclusions emerge:

- Luck was needed. For example, without the 45° tilt, uranium would not have been concentrated; the chemical composition of the rock did not poison the reactor; no geological process destroyed the reactor after it was formed.
- The Oklo reactor does not look like a man-made reactor, but it does share the principle of operation with the artificial, more elaborate reactor.

We can also see some parallels with the origin of life:

- There were stringent geological requirements for the origin of life. Without liquid water, and other conditions favourable for the appearance of organic compounds, life would not have emerged.
- Chemical cycles, formally acting as catalysts, are basic to life. Naturally occurring catalyst poisons could not have been present at the site of life's origin.
- Primitive life may have possessed mechanisms not present in extant organisms, but the basic principle of operation must have been present.
- The first biological systems could have been of very low efficiency in comparison with present ones.

2.3 The chemoton

Gánti (1974, 1975, 1979, 1987) proposed a model, the chemoton, which grows by metabolism, reproduces by fission in the biological sense, and which has rudimentary hereditary variation (Fig. 2.2). The chemoton consists of three subsystems: an autocatalytic network for metabolism; a bilayer membrane; and a replicable information carrier molecule. We discuss these functions in turn.

The metabolic network is an autocatalytic non-enzymatic chemical cycle. The cycle depicted consumes X as a raw material (X is supposed to be generated by the abiotic processes discussed below), and produces the following molecules:

Y as waste;

V', the monomer of the genetic material;

T', the precursor of the membranogenic molecule T; and two A_1 molecules for the original one A_1.

It is this last step that makes the cycle as a whole autocatalytic: the system catalyses its own production.

We have more to say about the plausibility of autocatalytic cycles in the next chapter. For the present, it is important to emphasize that autocatalysis is not the same as replication. For replication, it is not sufficient that an A gives rise to two As: it is also necessary that, if the A is replaced by a B (or a C or a D), then the

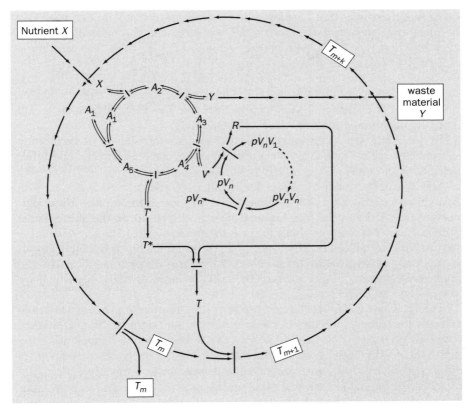

Fig. 2.2 The chemoton (Gánti, 1984). The metabolic subsystem, with intermediates A_i, is an autocatalytic chemical cycle, consuming X as nutrient and producing Y as waste material; pV_n is a polymer of n molecules of V', which undergoes template replication; R is a condensation byproduct of this replication, needed to turn T' into T, the membranogenic molecule; the symbol T_m represents a bilayer membrane composed of m units made of T molecules. It can be shown that such a system can grow and divide spontaneously.

cycle should give rise to two *B*s (or *C*s, or *D*s). In autocatalysis, there is no variation, and hence no heredity. Autocatalysis is an important first step towards replication, but it is not the whole road.

No-one, of course, thinks that such a small cycle would produce so many diverse compounds. In the living world, autocatalytic cycles are much more complicated, in three ways: (1) each reaction is catalysed by an enzyme; (2) the number of reactions is large; and (3) usually, there are networks rather than cycles. Examples are the CO_2-fixing Calvin cycle in plants, and the reductive tricarboxylic cycle in certain bacteria. These autocatalytic processes serve as the chemical basis for biological growth. Nevertheless, we shall keep to the idealized version for the time being.

The model assumes that T', after a simple reaction to give T, inserts itself into the membrane spontaneously, for the following reason. The molecule T is amphipathic (that is, it has a hydrophobic and a hydrophilic pole). The membrane is not a rigid structure: small gaps open and close in it from time to time. The hydrophobic tail of an arriving T molecule would slide into such a gap, since it is energetically more comfortable inside the bilayer: the reverse reaction would be rare, since it is energetically unfavourable. Hence the autocatalytic cycle will make the membrane grow, by the production of T'.

The preceding argument makes an important assumption about membrane permeability: while X and Y diffuse through the membrane freely, the other molecules cannot do so. We return to this question later.

There is another problem. When T molecules are added to the inside layer of the membrane, this layer will become denser relative to the outside layer: ultimately this will distort the membrane. In living cells, this distortion is prevented by the action of flippase enzymes that catalyse the so-called flip-flop reaction—the tumbling of molecules from one layer to the other. Since this involves an energetically unfavourable intermediate, it requires enzyme catalysis. In the absence of enzymes, membrane growth would be a peculiar process, which we now consider in more detail.

Initially, water will enter the microsphere, until the tension of the membrane balances the osmotic pressure of the internal solution: an equilibrium spherical state will arise. We now apply Koch's (1985) ideas to the behaviour of the microsphere. The activity of the autocatalytic cycle will produce compounds within the microsphere, with a concomitant increase in osmotic pressure. The stretching of the membrane will facilitate the incorporation of Ts into its inner layer, but, in the absence of enzymes, these molecules, instead of jumping into the outer layer, will form an inner membrane by invagination (Fig. 2.3). Meanwhile, some molecules do manage to jump, increasing the volume somewhat. Most molecules will jump at the point where the septum originates, because energetic conditions are more relaxed there. However, the volume of the compartment can only increase when, driven by increasing osmotic pressure, Ts jump to the outer layer in sufficient numbers. Koch (1985) calculated that spontaneous division is energetically possible. In fact, mycoplasmas and wall-less bacteria (L-forms) may follow this division mechanism rather closely. We return to this problem of cell division in Chapter 7.

We conclude that Gánti's microsphere grows by metabolism, and reproduces by fission in the biological sense. But what of hereditary variation?

pV_n is the symbol for the replicative template: it consists of a polymer of n molecules of type V. The model is so arranged that growth of the membrane occurs only if the template is replicated: thus the membranogenic T is formed from T' only in the presence of R, which is a byproduct of replication. The latter requires the opening of the two strands of pV_n and the formation of new ones by polymerization of V'. If the concentration of V' is too low, this process will not be

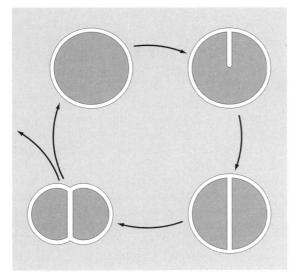

Fig. 2.3 Cell division driven by internal biosynthesis of phospholipids. The dimensions of the lipid bilayer have been exaggerated.

completed. Since an increase in volume would decrease the concentration, there is a coupling between replication and membranogenesis.

There remains the question of how far the chemoton has heredity. Genetic change without sequentially encoded information is possible in the following way. Imagine that the template pV_n is replicated by homologous pairing of Vs to the template. If, during copying, one of the daughter strands becomes shorter or longer, this change will be inherited by one of the daughter chemotons. Since there is coupling between metabolism and template replication, the presence of a pV_{n-1} molecule will alter the dynamics of the metabolic cycle, and will therefore affect the generation time of the whole system. This conclusion is borne out by simulations (Gánti, 1979). The transmission of hereditary information will be increased if the polymer contains two kinds of polymer, V and Z. If the template is replaced by a pV_nZ_m molecule, information will be coded, not only by the numbers n and m, but by the ratio n/m. Although the sequence is not utilized, either in coding or catalysis, it is inherited, ready for use at some later stage. In more advanced chemoton versions, it is assumed that the sequence contributes to a catalytic function of the genetic material, in a manner analogous to RNA enzymes (ribozymes).

In order to grow and replicate, the chemoton depends on an autocatalytic cycle and an exogenous source of energy. The hard problem, of course, is that in existing organisms catalysis depends on enzymes, and these in turn depend on the genetic information carried by nucleic acids. Nevertheless, we think that the abstract system of Fig. 2.2 is an excellent mental jumping-board for understanding the origin of life. We now consider how the components of such a system could have emerged on the primitive Earth.

3 Chemical Evolution

3.1 Introduction

Even if we take a 'replicators first' view of the origin of life, we still have to explain how a chemical environment arose that was sufficiently diverse for such replicators to be formed. As a minimum, we must explain the abiotic formation of sugars and amino acids, and of nucleotides or of some simpler molecules capable of base pairing, and hence of template reproduction. Lipids would not be needed for the first appearance of replicating molecules, but were crucial if evolution was to proceed further. Their importance is apparent from the chemoton model: by forming membranes, they make possible the formation of protocells, and hence of a unit of selection more complex than a replicating molecule. In turn, this encourages the evolution of cooperation between replicators, as discussed in the next chapter.

This chapter describes how such a diverse chemical environment could arise. In section 3.2, we introduce the familiar idea of a 'primitive soup', and review the experimental evidence concerning what compounds can, and cannot, be formed. Although a surprising array of organic compounds have been synthesized in such experiments, there are some awkward gaps. Section 3.3 describes the more recent, and more promising, idea that the crucial chemical events took place between compounds bonded to a charged surface. There are several reasons why the idea is attractive. Chemical reactions will be both more frequent and more specific if the reacting molecules are moving on a surface, rather than in three dimensions. Also, some reactions will be energetically favoured. In particular, the formation of biological polymers (proteins, nucleic acids) requires the removal of a molecule of water: this is difficult for molecules in solution in water, but easier on a surface.

The chemoton model incorporates an autocatalytic cycle, producing two molecules of a compound where one existed before. Such cycles are discussed in section 3.4. They are an essential feature of metabolism. It is important, however, to remember that autocatalytic cycles are not replicators: this distinction is made clearer in Chapter 4.

Section 3.5 discusses whether chemical 'evolution', in the primitive soup or on a surface, is really evolution. In our view, evolution requires heritable variation: this property is lacking from the reactions discussed in this chapter. However, if metabolism did indeed originate on a surface, this has some intriguing implications for the emergence of units of selection more complex than single molecules. As we argue in Chapter 7, it makes it easier to see how protocells, bounded by lipid membranes, could have arisen. More immediately, it means that chemical reactions took place between molecules that were fairly permanent neighbours, instead of only momentarily in contact as would be the case in solution. This at once brings to mind the analogy with the role of kin selection in the evolution of cooperation between organisms that are neighbours. However, we argue in section 3.5 that this analogy is misleading, at least until

replicating molecules exist. Once replicators exist, interactions between neighbours do become important, as discussed in Chapter 4.5.

Even if, in the absence of replicators, there was no room for the operation of kin selection, interactions between neighbouring autocatalytic cycles could still have been important in generating greater chemical diversity. This possibility is discussed in section 3.6. Finally, in section 3.7 we describe two alternative sources of organic compounds—meteorites and chemical reactions in clouds. We see these as additional sources of diversity.

3.2 Experiments: the primitive soup

Oparin (1924) and Haldane (1929) independently suggested that, if the primeval atmosphere was reducing, and if there was a supply of energy—for example, from lightning or ultraviolet light—then a wide range of organic compounds might be synthesized. Miller (1953), on the advice of Harold Urey, tried to mimic conditions on the primitive Earth. Besides water, some other simple compounds must be present, as well as a source of energy (Fig. 3.1). At that time it was thought that the primitive atmosphere was reducing, so the gas mixture in the flask consisted of CH_4, NH_3 and H_2O. Electric discharges between two electrodes simulated ancient lightning. The results were astonishing: several proteinogenic amino acids were formed, as well as other organic compounds. Many similar experiments have since been conducted: Tables 3.1 and 3.2 list the coded and non-coded amino acids thus formed. It is encouraging that the amino acid composition of the Murchison meteorite is strikingly similar.

Given the existence of amino acids, what of the formation of polypeptides? Fox

Table 3.1 Protein amino acids from the Miller–Urey experiment

Amino acid	$CH_4/N_2/NH_4Cl^a$ (1:1:0.05 mol)	$CO/N_2/H_2^b$ (1:1:3)	$CO_2/H_2/N_2^b$ (1:3:1)
Glycine	100	100	100
Alanine	180	2.4	0.87
Valine	4.4	0.005	<0.001
Leucine	2.6	—	—
Isoleucine	1.1	—	—
Proline	0.3	—	—
Aspartic	7.7	0.09	0.14
Glutamic	1.7	0.01	<0.001
Serine	1.1	0.15	0.23
Threonine	0.2	—	—

Mole ratios normalized to glycine as 100.
From Miller, 1987.

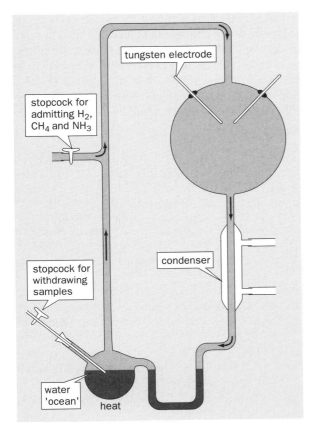

tungsten electrode

stopcock for admitting H_2, CH_4 and NH_3

Fig. 3.1 The Miller–Urey reactor for prebiotic organic synthesis.

condenser

stopcock for withdrawing samples

water 'ocean' heat

Table 3.2 Non-protein amino acids from the Miller–Urey experiment

Amino acid	$CH_4/N_2/NH_4Cl$	Amino acid	$CH_4/N_2/NH_4Cl$
Sarcosine	12.5	N-Methyl-β-alanine	1.0
N-Ethylglycine	6.8	N-Ethyl-β-alanine	0.5
N-Propylglycine	0.5	Pipecolic	0.01
N-Isopropylglycine	0.5	α-Hydroxy-γ-aminobutyric	17
N-Methylalanine	3.4	α, β-Diaminobutyric	7.6
N-Ethylalanine	trace	α, β-Diaminopropionic	1.5
β-Alanine	4.3	Isoserine	1.2
α-Amino-n-butyric	61	Norvaline	14
α-Aminoisobutyric	7	Isovaline	1
β-Amino-n-butyric	0.1	Norleucine	1.4
β-Aminoisobutyric	0.1	Allothreonine	0.2
γ-Aminobutyric	0.5		

From Miller, 1987.

& Dose (1977) kept a mixture of amino acids, with an excess of aspartic and glutamic acid, for 7 days under anhydrous conditions at 120 °C, and then put the mixture into water. The result was the appearance of proteinoids, or thermal proteins, with weak catalytic activity. Given the existence of volcanoes, the physical conditions are not implausible. Fox (1984) has emphasized that the sequences formed are non-random. However, this does not make much difference: sequences with heritable variation would be exciting, but mere non-randomness is not. Perhaps more important, the amino acids are linked by a variety of chemical bonds, and not only by peptide bonds as in biological proteins.

The abiogenic synthesis of nucleic acids requires first that of the nucleotides. Various purines and pyrimidines, including A, G, C and U (but no T), have been synthesized in such experiments. HCN seems to be a key compound in these processes: in fact, A can be thought of as an HCN polymer (Fig. 3.2).

The origin of sugars, including ribose, seems readily explicable by the prebiotic functioning of the formose reaction, first described by Butlerow in 1861. In fact, we are dealing here with a complex network of reactions, producing sugars from pre-existing sugars and formaldehyde (Fig. 3.3). Once the first intermediate,

Fig. 3.2 Adenine as an HCN polymer.

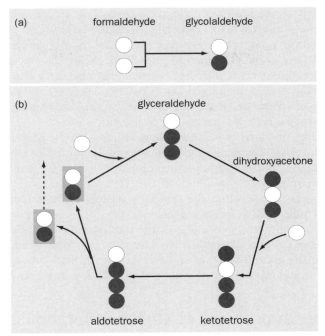

Fig. 3.3 The autocatalytic core of Butlerow's (1861) 'formose reaction'. (a) Spontaneous, slow formation of glycolaldehyde from formaldehyde. (b) After one turn of the cycle, one new molecule of glycolaldehyde is produced. The structural isomers of sugars $(CH_2O)_n$ are specified by the carbon skeleton and by the position of the carbonyl $(C = O)$ group (open circle).

glycolaldehyde, is formed by a slow reaction uniting two formaldehyde molecules, the network begins to work at an accelerating rate, since the synthesis of these sugars is autocatalytic.

There are two problems with this network that should be mentioned. First, the sugars formed are rather unstable, so, if they are to be present in significant amounts, this can only be in a steady state of formation and decay. It is imperative, therefore, that the end products of sugar decay be recycled to formaldehyde. Second, it is not at all obvious how ribose, among the more than 40 sugars, could have been sufficiently prevalent under prebiotic conditions.

Setting aside the problem of the origin of ribose, the synthesis of nucleosides (base and sugar linked together as in present-day nucleotides) also poses problems. Purines react with ribose to yield the corresponding nucleosides in small amounts. The analogous reaction with pyrimidines seems hopeless.

The phosphorylation of nucleosides to nucleotides can be done in dry-phase with relatively good yield, but all sorts of isomers with varying degrees of phosphorylation emerge. This lack of purity is important because accurate re-

plication of a polymer depends on chemical purity. We return to the formation and replication of nucleic acids, especially RNA, in section 5.4.

Lipid formation again could have been preceded by the appearance of their constituents: fatty acids, glycerol and phosphate. While the abiogenic reaction between these three seems plausible, we have trouble with the formation of membranogenic lipids: no long-chain (C6–C18) linear (non-branched) fatty acids have been synthesized in electric discharge reactions, although they would be indispensible for prebiotic membrane formation.

Summarizing, one is left with ambivalent feelings. On the positive side, one is amazed by the ready formation of several biologically significant compounds, but it is discouraging that many important molecules resist prebiotic synthesis in acceptable quantities.

Finally, the assumption that the primitive atmosphere was reducing has recently been challenged. If, as we suggest in section 3.3, the earliest stages of life occurred on the surface of pyrite (FeS_2), the nature of the atmosphere is largely irrelevant, since the crucial events took place close to hydrothermal vents in the deep sea. But the accumulation of organic compounds in clouds, as discussed in section 3.7, would have required a neutral or reducing atmosphere.

3.3 The hypothesis of surface metabolism: the primitive pizza

An alternative and in some ways more promising picture of chemical evolution has been suggested by Wächtershäuser (1988). The reactions he suggests are so specific that it should be possible to test them in the laboratory. His central idea is that chemical evolution, leading to the appearance of life, started with a surface-bonded autocatalytic chemical network, consisting of polyanionic organic compounds. There are three aspects of this idea: (1) the thermodynamics; (2) the kinetics; and (3) the chemical feasibility of surface bonding. We shall discuss these in turn.

The essential point about entropy is as follows. If A and B combine to form a complex AB in solution, the result is a great reduction in the freedom of motion of A and B, and hence of the entropy of the system. In contrast, if A and B are surface-bound, they have a much reduced freedom to move around: they may slowly walk on the surface in two dimensions, and rotate around one axis only. Therefore, the decrease in entropy is not particularly high, which favours the reaction. Entropy considerations also show that the formation of peptides from amino acids is favoured if the peptides are bound to a surface. Peptide formation involves the elimination of a molecule of water: on a surface the elimination of a water molecule increases the entropy, and so the reaction is more likely.

The kinetic advantages of surface binding are obvious. For example, many extant enzymatic reactions require the successful collision of three molecules. Such collisions are extremely unlikely in solution, but less so on a surface.

How feasible is surface bonding? The binding must be strong, but flexible enough to allow slow migration in two dimensions on the surface: only the ionic bond satisfies these criteria.

The usual cations in biochemistry are organic nitrogen bases. However, no insoluble salts of these compounds are known. This leaves us with the possibility that the mineral surface should be positive, and the anchors of the organic molecules should be anionic. Rather surprisingly, this eliminates clay from the candidate surfaces. Wächtershäuser favours pyrite (fool's gold) as the primary site for surface metabolism, since it is an insoluble mineral with a positive surface charge.

Chemistry suggests that the organic anions should have at least one group with at least two negative charges (such as $-PO_3^{2-}$) or at least two groups with at least one negative charge (such as $-COO^-$), to ensure insoluble binding to pyrite. This implies that those molecules of extant intermediate metabolism that have such anchors may be very ancient: similarly, surface bonding may be the most ancient function of phosphate groups.

3.4 A logical basis for autocatalysis

In the chemoton, an autocatalytic cycle is an essential component. It is important to understand why. Suppose that the cycle in Fig. 2.2 was not autocatalytic: that is, no second molecule of A_1 was produced. Starting with 100 molecules of A_1, one operation of the cycle would produce 100 molecules each of T' and V', which are needed for the growth of the system, and also 100 molecules of A_1, which would serve as the starting point of the next cycle. What is wrong with this idea is that it assumes perfect specificity of all the reactions. In practice, all sorts of side reactions would occur. Starting with 100 A_1 molecules, you would be lucky to finish with 90: the cycle would rapidly run down. Autocatalysis—the production of two molecules of A_1 from one—is needed to compensate for errors.

Figure 3.4 illustrates three cycles, in abstract form. Only the last two are autocatalytic. How plausible are such cycles? King (1982) has shown that they are more likely than one might think. His basic conclusion is that the sufficient conditions for autocatalysis in a recycling network are:

- a branching subsystem that can be cyclically closed without the intermediates (e.g. A_1 and A_3 in Fig. 3.4c) reacting; and
- the network should contain at least two bimolecular reactions (Fig. 3.4 contains three such reactions).

These conditions are not severe: autocatalysis is quite likely in recycling systems.

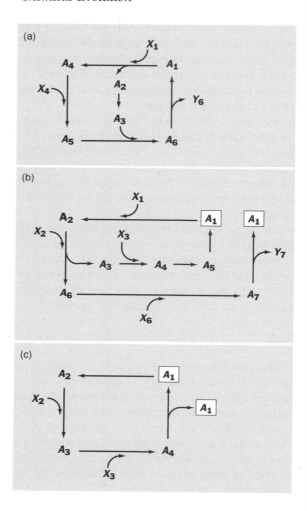

Fig. 3.4 Three different chemical cycles in abstract form. (a) This cycle is not autocatalytic. (b) An asymmetrically autocatalytic cycle. (c) A symmetrical autocatalytic cycle. A_i cycle intermediate; X_i raw material; Y_i waste.

3.5 Is chemical 'evolution' evolution?

Gánti (1974, 1979) was the first to call attention to autocatalytic cycles from the point of view of evolution. He pointed out that if two cycles are operating in the same environment, then the one with the larger kinetic rate constant can outgrow the other. This process is analogous to the competitive replacement of one species by another. As in the ecological context (Gause, 1934), one cycle will replace the other only if they are limited by the same nutrients, supplied from outside the cycle: if two cycles depend on different nutrients, they may coexist. We discuss this possibility in section 3.6.

For evolution, however, we require more than autocatalysis. We require that, occasionally, a new variant chemical compound—a mutation—should arise, and, once arisen, should be replicated. In the case of autocatalysis, this would require that, in a cycle such as Fig. 3.4c, if A_1 were replaced by a different molecule, A_z, the result would be the production of more A_z: this, of course, is usually not so, although Wächtershäuser (1992) has given some hypothetical but chemically plausible examples.

Is there, then, any significance from an evolutionary point of view in the fact that chemical reactions occurred on a surface? In the evolution of higher organisms, if offspring are near to their parents, or, more generally, if related individuals interact with one another, this favours the evolution of cooperation by kin selection. But this depends on hereditary replication. Once replicating molecules existed, relatedness between neighbours, arising because of limited movement, could have been important: we return to this point in section 4.5. But prior to the origin of replication, relatedness is a meaningless concept.

3.6 Evolution of metabolic networks through chemical symbiosis

King (1982, 1986) pointed out a real difficulty about increasing chemical complexity in the primitive soup. Imagine an autocatalytic cycle of n steps. Suppose that, in the i^{th} step, the probability that the intermediate a_i will be converted into the next step in the cycle, and not be shunted off by some side-reaction, is p_i. Then the gain in the cycle is $g = 2 \cdot p_1 \cdot p_2 \cdots p_n$, or, if all ps are equal, $g = 2p^n$. Hence the condition for the growth of an autocatalytic cycle is

$$p^n > 0.5. \tag{3.1}$$

Specificity of the reaction, therefore, is crucial. More specific reactions permit longer cycles. For a living cell, high specificity is ensured by the selectivity of enzymes, and the shielding effect of the cell membrane, neither of which were available for abiogenically produced autocatalytic cycles. The relatively smaller number of chemical species present in surface metabolism, compared to the chemical chaos of the primeval soup, would favour specificity, but the problem of specificity remains a serious one.

Even if we accept that surface metabolism could satisfy condition 3.1, there remains the problem of further increase in complexity. Happily, King's (1980) idea of chemical symbiosis between autocatalytic cycles may save the situation. Imagine two cycles, A and B, requiring among their nutrients X_A and X_B, and producing wastes Y_A and Y_B. Let $X_A = Y_B$ and $X_B = Y_A$: that is, the waste product of each cycle is the nutrient of the other. This makes a stoichiometric coupling, a chemical symbiosis, between the two cycles.

In surface metabolism, this would happen when two domains, occupied by cycles *A* and *B*, become adjacent: symbiosis might be established. Interactions between neighbours, therefore, although irrelevant to evolution until replicators existed, could have been important in the maintenance of chemical diversity.

3.7 Chemical evolution in clouds, and the extraterrestrial contribution

Woese (1979) proposed another alternative to the primitive soup: he suggested that chemical evolution took place on the surface of water droplets in clouds. He had two main reasons:

- The temperature of the surface of the Earth may have been too high to permit the existence of oceans. More recently, Kasting & Ackerman (1986) estimated that the surface temperature of the early Earth would not have exceeded 100 °C. However, it is still possible that frequent collisions by meteorites (Maher & Stevenson, 1988) ensured that the oceans did not last long enough for chemical concentrations to reach a high value. But this need not have mattered, because liquid water would still be present in droplets in the atmosphere, and it may be easier to concentrate organics in droplets than to turn the ocean into a primitive soup.

- Water drops in clouds would provide a vast surface area for chemical reactions. Further, water droplets frequently form around solid particles. Since polymerization is a condensation reaction, the alternate forming and drying of drops would favour polymerization, as in Fox's experimental formation of proteinoids described above.

The idea is attractive, but, like Wächtershäuser's idea of surface metabolism, it is a mechanism for the production of a diverse chemical environment, not for the origin of replicating entities. Thus, water droplets are not primitive cells—they do not divide. Their importance is that they provide a surface for chemical reactions. Oberbeck *et al.* (1991) have worked out further details of such a chemistry.

As a final contribution to prebiotic chemistry, there is the possibility, first suggested by Oró (1961), that comets and meteorites delivered important compounds to the early Earth. The extent of this delivery depends on the characteristics of the early atmosphere. The present-day atmospheric pressure would not be sufficient to prevent impact from meteorites as small as 100 m in radius, with velocities below 10 km/s. Interactions with the atmosphere, and surface impact, would destroy any organic compounds present. But if we assume a more plausible 10-bar CO_2 atmosphere, then 4500 million years ago the Earth was collecting cometary organics at a rate greater than 10^6–10^7 kg/year (Chyba *et al.*, 1990).

3.8 Conclusions

This chapter has been concerned with the establishment of a chemically diverse environment, in which the first polymers capable of template replication could have arisen. Experimental investigation of the Haldane–Oparin idea of a primitive soup has shown that a wide range of compounds can be produced by very simple means. But difficulties remain, particularly in the formation of ribose and pyrimidines, and in polymerization in aqueous solution. The idea of surface metabolism offers a possible solution to these problems. The need now is for experimental studies of surface metabolism, similar to the earlier studies of the Haldane–Oparin hypothesis.

There is no contradiction between the theories of cloud chemistry and surface metabolism. Organics synthesized in clouds and washed into the ocean could have been captured by pyrite. Pyrite itself could have been formed within atmospheric droplets by the reactions around the dust-grain nuclei. The same pyrite particle could have spent some of its time in the ocean and some in the atmosphere, preserving its precious inhabitant, the primordial pizza.

4 The Evolution of Templates

4.1 Introduction

In this chapter, we discuss the origin and early evolution of genetic replication. The argument is complex, so we start with a brief outline. Section 4.2 discusses the nature of replication. We draw a distinction between simple replicators, limited hereditary replicators and indefinite hereditary replicators. Continued evolution requires indefinite hereditary replicators: it seems that such replicators depend on some form of template reproduction. In section 4.3, we point out that there is an error threshold for the accuracy of replication: for a given total quantity of genetic information—for example, for a fixed number of bases—there is an upper limit on the error rate of replication. If the error rate rises above this limit, natural selection cannot maintain the information. This leads to what we have called Eigen's paradox. In the absence of specific enzymes, replication accuracy is low. Hence the total genome size must be small—almost certainly, less than 100 nucleotides. The genome is therefore too small to code for accurate replication machinery. There is a catch-22 situation: no enzymes without a large genome, and no large genome without enzymes.

The next three sections discuss possible solutions to the paradox. Section 4.4 considers populations of replicating RNA molecules. We point out that the dynamics of replication are such as to lead to the stable coexistence of a diverse population, but we do not think that this constitutes a solution to the paradox. Section 4.5 discusses the hypercycle, a particular relationship between replicators that makes it possible for a greater total quantity of information to be maintained with a given accuracy of replication. We argue that the further evolution of hypercycles requires that they be enclosed within compartments, because otherwise they are sensitive to parasitic replicators. We also discuss, rather inconclusively, the possibility that, even in the absence of compartments, cooperation might evolve, by a processes analogous to kin selection, if the components of the hypercycle were confined to a surface. Finally, we discuss an alternative model, the stochastic corrector model. This also depends on the existence of compartments, but emphasizes the importance of stochastic effects arising if there are small numbers of each kind of molecule in a compartment. Essentially, small numbers serve to generate variation upon which selection can act.

4.2 Replication and nucleation

The term 'replicator' has several different senses. It may mean a structure that can arise only if there is a preexisting structure of the same kind in the vicinity. Such structures we will call 'simple replicators': as an example, peroxisomes, mentioned in Chapter 8, appear to be simple replicators. Some replicators, however, are able to exist in several forms—A, B, C, etc.—and, when they

replicate, the new structure resembles the old. We will call such structures 'hereditary replicators'. A DNA molecule is such a hereditary replicator, and the number of possible distinct replicable structures is indefinitely large: DNA molecules, therefore, are 'indefinite hereditary replicators'. It is possible, however, that there are also structures which can exist in a limited number of heritable states: these we will call 'limited hereditary replicators'. We shall meet this possibility when discussing the evolution of microtubular structures in Chapter 8.

It is a familiar fact that a supersaturated solution will crystallize if a small seeding crystal is added. A model of this process of nucleation is shown in Fig. 4.1. If we suppose in addition that, when a polymer reaches a given length, it breaks in two, then we have a population of 'simple replicators', as defined above. However, only one type of replicator is possible. Figure 4.2 shows a slight modification of the scheme. If the shape of the monomers is slightly changed, two different types of polymer, or of replicator, are possible, and each will breed true: these are limited hereditary replicators. The more complex system, of indefinite hereditary replicators, is familiar to biologists because it is represented by nucleic acid replication. Replicators of the first two kinds can be illustrated by simple plywood models (Penrose, 1959).

One final question is as follows: given that there are two potential replicators, as in Fig. 4.2, can nuclear genes influence which one will be realized? The monomers α and β represent proteins whose shape is determined by genes. It would be possible for genes also to determine a self-assembling scaffold, which

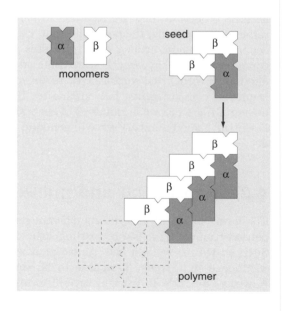

Fig. 4.1 Nucleation. Suppose that two monomers will join only if two 'bonds' are formed between them. Then free monomers will not join, but if a seed is added, it will bind monomers to form a polymer.

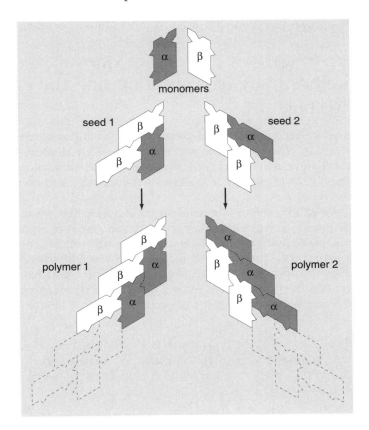

Fig. 4.2 Limited hereditary replicators. The same two monomers, α and β, can form two different polymers, depending on the nature of the initial seed (the rules for joining are as in Fig. 4.1).

acted as a template for the formation of one type of seed or the other. Hence the answer to the question is yes.

These models show that entities can be replicators without having variation or heredity, and that they can have variation and heredity limited to relatively few types. For sustained evolution, however, we require indefinite hereditary replicators. This has only been achieved by template reproduction involving complementary base pairing.

The first artificial replicator, not needing enzymes for its replication, was synthesized by von Kiedrowski (1986), on the basis of careful chemical considerations. By now, a family of such molecules are available (reviewed by Rebek, 1994). Although by their nature they are not immediately relevant to the origin of life, they demonstrate the existence of limited hereditary replicators. The

classification of replicators is further discussed by Szathmáry & Maynard Smith (1993b).

4.3 The accuracy of replication and the error threshold

The first important experimental results on molecular evolution are described by Spiegelman (1970). The radical step was to replicate RNA *in vitro*. This needs nothing more than an RNA template, activated ribonucleotides, and a replicase enzyme from an RNA phage, $Q\beta$. The method of serial passage is shown in Fig. 4.3.

The wild-type phage consists of more than 4000 nucleotides. If this was used as the initial primer, it was found that after several rounds of replication the survivor RNA molecules were only a few hundred nucleotides long. The explanation is as follows. Variant RNAs, including broken and deleted mutants, appear during replication. Those molecules that replicate fastest have the best chance to get into the passaged sample, and to continue replicating in the next test tube. The only fitness criterion is replication speed. Since shorter molecules

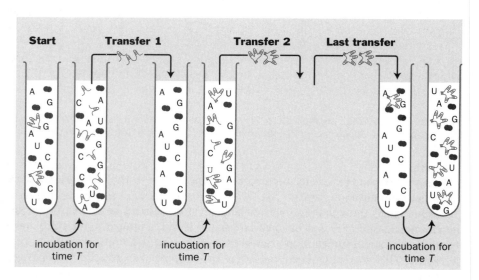

Fig. 4.3 Evolution *in vitro*. The initial tube contains the four nucleotides from which RNA is synthesized, the replicase enzyme from the $Q\beta$ phage and a 'primer' RNA molecule. This molecule is copied many times, with some errors. After incubation for time *T*, a drop of solution is transferred to a new tube containing nucleotides and replicase (but not, of course, any primer). The process of incubation and transfer can be repeated indefinitely, and evolutionary changes in the population of RNA molecules can be observed.

replicate faster, they are selected for. Why does the length of the RNA not decrease still further, say to 50 nucleotides? The answer lies in the relation between the structure of RNA and its replicability. RNA molecules form self-paired and quasi-globular structures, similar to tRNA molecules, which are re-cognized by the replicase before it starts work. Too short a molecule would not be recognized, and so would not be replicated.

Which is the fittest RNA molecule depends on the environment. The addition of inhibitory chemicals to the incubating solution prohibits template replication of all but a few RNAs, which can be copied at a very low speed. The descendents of the latter may become, through mutation, more resistant to the inhibitor. In practice, this is what happened: the wild-type replication rate was almost completely restored, as a result of three point mutations in the original RNA. Other RNAs became not only resistant to but also dependent on the inhibitor: without it, they replicate more slowly.

Manfred Eigen and his colleagues have developed a formalism to describe the mutation-selection processes of RNA molecules. This has led (Eigen, 1971) to the discovery of a paradox of prebiotic evolution, which we now describe. For simplicity, we disregard the complementary nature of replication, and proceed as if a direct copy were made of the template. Let the copying fidelity of one base be q, where $0 < q < 1$. Then the error rate per base per replication is $(1 - q)$. The copying fidelity of a template N nucleotides long is $Q = q^N$, if we assume that copying errors occur independently. Since N figures as an exponent, a large N gives a small Q, even if q is close to 1.

Imagine a population of molecules in which each unique sequence is assigned a number, $i = 1$ to n. The concentration of molecules with sequence i is x_i. Then the rate of change in the concentration x_i is:

$$\dot{x}_i = (A_i Q - E)x_i + \sum_{j=1}^{n} w_{ij}x_j. \qquad (4.1)$$

In this Eigen equation, A_i is the total rate of copying of sequence i, irrespective of fidelity; w_{ij} is the mutation rate, giving the probability of the production of sequence i by miscopying of j; E is a variable introduced to ensure that the total concentration remains constant: this constraint imposes selection on the mole-cular population. For simplicity, we assume that all molecules are the same length, and hence that Q is constant.

If we ignore the second term, measuring the effect of mutation, the sequence with the highest $A_i Q$ value will displace all others: thus fitness is the product of two factors—overall copying rate and copying fidelity. In general, however, we cannot ignore mutation. This will be analysed under the following simplifying assumption (Maynard Smith, 1983a). We divide the population into two sub-populations: the fittest or master sequence (number m), and all the mutants. The latter are replaced by an average sequence (number j). We can ignore back

mutations to the master, because a mutant sequence is enormously more likely to mutate to another mutant than back to the master. Then the equations become:

$$\dot{x}_m = A_m Q_m x_m - x_m(A_m x_m + A_j x_j), \qquad (4.2a)$$

$$\dot{x}_j = A_j x_j + A_m(1 - Q)x_m - x_j(A_m x_m + A_j x_j), \qquad (4.2b)$$

where

$$x_m + x_j = 1,$$

which is the selection constraint: one sequence can grow only at another's expense. The last factor in parentheses in both equations corresponds to E in equation 4.1, introduced to ensure a constant total concentration.

We are interested in the coexistence of master and mutants. The alternative is that only mutants are present at equilibrium: selection is not able to maintain the master sequence against mutational decay. We therefore set both rates of change to zero, and ask whether an equilibrium with non-zero x_m can exist. It turns out that such an equilibrium requires

$$Q > A_j/A_m = 1/s, \qquad (4.4)$$

where s is the selective superiority of the master. We know, however, that

$$Q = q^N \approx e^{-N(1-q)}. \qquad (4.5)$$

Combining the last two equations we have

$$N < \ln s/(1 - q), \qquad (4.6)$$

which means that the selectively maintainable amount of information (N) is limited by the copying fidelity per digit (q) (Eigen, 1971).

The relationship described by equation 4.6 is shown in Fig. 4.4. The more accurate the replication, the longer the selectively maintainable master copy. If the master grew beyond the error threshold, it would quickly deteriorate. Below the error threshold, the population would consist of the master sequence, surrounded by a cloud of its closest mutants: this molecular distribution has been called a quasispecies (Eigen & Schuster, 1977).

Measurements on the $Q\beta$ phage and foot-and-mouth disease virus are consistent with this model (Domingo *et al.*, 1980). The mutation rate per base of $Q\beta$ is $1 - q = 5 \times 10^{-4}$: hence, this virus approaches its error threshold, with $N = 4500$.

If the error threshold places an upper limit on genome size, we would expect an inverse relationship between genome size and mutation rate per base. At a crude level, this certainly holds; for example, RNA viruses have inaccurate replication and small genomes. Drake (1991) found a striking relationship for a

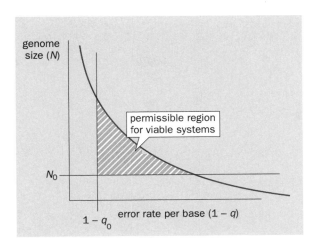

Fig. 4.4 Permissible genome size as a function of replication accuracy. N is the number of base pairs and q the probability that a single base is correctly replicated. The hyperbola is drawn for ln $s = 1$, where s is the selective advantage of the 'master' copy over its mutants. Viable systems must lie below the hyperbola (text equation 4.6). There are other constraints. Genome size must exceed some threshold N_0, to encode sufficient information; the error rate cannot be reduced below some threshold $1 - q_0$ without excessive costs in replication time and energy. Hence, the permissible region is the hatched one.

Table 4.1 Rates of spontaneous mutation in DNA-based microbes

Organism	Genome size (bp)	Target	Mutation rate Per bp (μ_{bp})	Per genome (μ_g)
Bacteriophage M13	6.41×10^3	*lacZα*	7.2×10^{-7}	0.0046
Bacteriophage λ	4.85×10^4	*cI*	7.7×10^{-8}	0.0038
Bacteriophage T2	1.60×10^5	*rII*	2.7×10^{-8}	0.0043
Bacteriophage T4	1.66×10^5	*rII*	2.0×10^{-8}	0.0033
Escherichia coli	4.70×10^6	*lacI*	4.1×10^{-10}	0.0019
			6.9×10^{-10}	0.0033
		his GDCBHAFE	5.1×10^{-10}	0.0024
Saccharomyces cerevisiae	1.38×10^7	*URA3*	2.8×10^{-10}	0.0038
		SUP4	(7.9×10^{-9})	(0.11)
		CANI	1.7×10^{-10}	0.0024
Neurospora crassa	4.19×10^7	*ad-3AB*	4.5×10^{-11}	0.0019
		mtr	(4.6×10^{-10})	(0.019)
			1.0×10^{-10}	0.0042

range of organisms from DNA phages, through prokaryotes to the simplest eukaryotes (Table 4.1). Yet it is hard to believe that this reflects the error threshold, for two reasons. First, the mutation rate per genome, per generation is very low—an average of 0.0033: the error threshold would permit a much

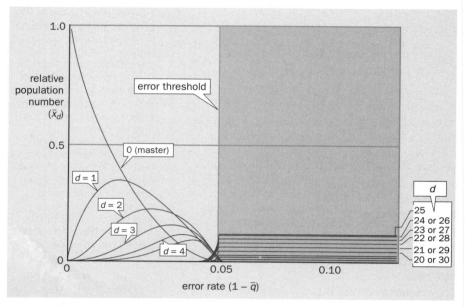

Fig. 4.5 The error threshold (Swetina & Schuster, 1982). *d* is the number of differences between a particular sequence and the 'master' sequence that has the highest fitness. Thus the curve for *d* = 3 is the proportion of sequences differing from the master by exactly three mutations. In this simulation, the sequence was of 50 sites, each occupied by one of two kinds of base. The fitnesses of all sequences other than the master were equal. Beyond the threshold, all sequences were equally frequent. Since the numbers of sequences with *d* = 23 and *d* = 27, for example, are the same, the frequencies of the two classes were also equal, as shown here.

higher value. Second, much higher values are in fact observed in higher eukaryotes. For example, the mutation rate per cell generation in mammals, although not known with the same accuracy as that of *E. coli*, is probably not greatly different. Yet the genome size of mammals is several orders of magnitude greater, and there are a number of cell generations per individual lifetime. We suspect that the genome size of prokaryotes is limited because there is only a single origin of replication (section 8.4), and not by the error threshold: but, if so, the constant per genome rate illustrated in Table 4.1 remains unexplained.

It is interesting to see how the error threshold manifests itself for the original equation (4.1) with many sequences. Figure 4.5 shows the quasispecies distribution for $N = 50$ for different values of q. Sequences are classified into mutant classes: all sequences of k mutations from the master are members of class k. All values of replication rate, A_j, other than that assigned to the master, are equal, and less than the master's. The logarithmic plot especially highlights the catastrophic flavour of the transition, which reminds one of a phase transition (Swetina & Schuster, 1982).

One question that has been investigated for the $Q\beta$ replicase system is the nature of the 'fitness surface': is there, for given physical conditions, a single optimum to which evolution always proceeds, or are there many local optima on which the population may be trapped? Sumper & Luce (1975) found, surprisingly, that the $Q\beta$ replicase can synthesize RNA molecules *de novo*, without added templates. This happens because the enzyme has a high affinity for certain tetranucleotides. If mononucleotides happen to assemble on the surface of the enzyme in the appropriate order, there is a chance they may be ligated, even though a complementary template is absent. In such experiments, the final result was uniform across many repetitions: the so-called midivariant (MDV) established itself. In other words, there was a single optimum. It later turned out, however, that this single optimum is a feature of experiments carried out at high salt concentrations. Biebricher *et al.* (1981) found, in comparable experiments, that different optima were reached in different test tubes, or the same optimum was reached by very different routes, even if the initial conditions were the same. Thus, in their experiments they observed a rugged fitness landscape, with many local fitness optima.

For the origin of life, it is the q of non-enzymatic replication that is relevant. Measurements of the interactions between paired nucleotides show that this value cannot be greater than 0.99. Hence N cannot be appreciably greater than 100 (Eigen, 1971). To increase this number, a replicase enzyme is needed. The smallest genome able to code for such an enzyme, and for the necessary translating machinery, would contain a number of genes, whose total length would greatly exceed 100 nucleotides. If the genes were unlinked, they would compete with each other, and some would be eliminated: if they were linked, the genome size would be well beyond the error threshold.

This is the catch-22 of prebiotic evolution: no enzymes without a large genome, and no large genome without enzymes (Maynard Smith, 1983a). We shall refer to this as Eigen's paradox (Szathmáry, 1989a,b). We discuss serious attempts to solve it in the sections on the hypercycle, and on the stochastic corrector model. First, however, we point out that there are reasons to expect different species of RNA molecule to have coexisted in the absence of any special mechanisms.

4.4 The ecology and coexistence of RNA molecules

In this section, we discuss the ecology of populations of replicating molecules. Because of the details of the dynamics, such populations are likely to become diverse, but we will argue that this fact does not provide a resolution of Eigen's paradox.

Studies of the replication of short oligonucleotide analogues (von Kiedrowski, 1986; Zielinski & Orgel, 1987) revealed that the kinetics of replication were peculiar: the concentration of templates increased less than exponentially with time. This type of growth has important consequences for selection (also reviewed by von Kiedrowski, 1993).

The normal growth for replicating entities is exponential, that is

$$\mathrm{d}x/\mathrm{d}t = kx, \tag{4.7}$$

where k is the malthusian rate of increase. If we have two species with different rate constants, k_1 and k_2, both will increase exponentially. In a finite world, however, there must be some factor limiting growth. If both species are limited by the same factor, then only the species with the higher rate constant will survive: the other will go extinct.

Things are very different if growth is subexponential. It turns out that, even if the two species are limited by the same resource, there is stable coexistence (Szathmáry & Gladkih, 1989). At first sight, this seems to contradict Gause's principle, that two species limited by the same resource cannot coexist. The contradiction, however, is only apparent. Subexponential growth implies that the growth of each species is also limited by its own numbers. The reason is as follows. Replication involves the production of a complementary strand. As concentrations increase, a greater proportion of the templates are paired with their homologues, and are therefore not available for replication. Note that this is a process in which the growth of each species is limited by its own concentration, but not by that of its competitors.

Hence, if replication is by complementary base pairing, we would expect a population of replicators to consist of many species, and not just of the most efficient species, even though the replicators are competing for the same monomers. Why, then, did Spiegelman (1970) observe a single optimal sequence, rather that a polymorphic population? The answer lies in the conditions of the experiment, which involved an excess of replicase enzyme. In the non-enzymatic case, double-strand formation is very effective. The $Q\beta$ enzyme, however, can unwind the duplex: hence, in Spiegelman's experiments, self-inhibition was greatly reduced. If template concentration is so high that it saturates the enzyme, however, duplex formation is substantial, and polymorphic molecular populations result (Biebricher *et al.*, 1985).

This phenomenon, 'the survival of everybody' (Szathmáry, 1991a), may have contributed to the diversity of early populations of replicators, but it does not invalidate Eigen's paradox. First, in the presence even of low-efficiency replicases, coexistence is possible only at high concentrations of templates, sufficient to saturate the enzyme: but then coexistence is at the expense of double strands that, in effect, are lost for ever. It is hard to imagine that, given such a system, a superior competitor would not have evolved. Second, as the diversity of

sequences increased, heterologous double strands would form, compromising stable coexistence. We conclude that subexponential growth, although permitting greater sequence diversity, is not an adequate resolution of Eigen's paradox: it would not allow the evolution of an entity with a greater total amount of genetic information than that permitted by the error threshold.

4.5 The hypercycle

Eigen himself has endeavoured to resolve his paradox. Since, because of the error threshold, it is impossible to store all the necessary information in a single giant chromosome, stable coexistence of several cooperating information carriers must be achieved. Some interaction other than competition between genes is needed. Eigen & Schuster (1977) have summarized the criteria for the integration of information as follows:

- The individual RNA molecules must still be able to compete successfully with their own erroneous copies.

- The different genes, serving different functions, should not competitively exclude one another.

- The new, integrated system must compete successfully against other less-efficient systems or individual sequences.

Eigen (1971) argues that some functional interaction between the different RNAs is necessary: this interaction arises, in his view, because RNAs have some replicase activity. But a common replicase copying every member of the system does not solve the problem: competitive tendencies would, as in Spiegelman's experiments, remain effective. Instead, the first member should specifically replicate the second, the second the third, the third the fourth, and so on, and the last member should catalyse the replication of the first: this is the hypercycle (Fig. 4.6). It is intuitively clear (and can be checked mathematically—see Szathmáry, 1989a,b for a simple derivation) that no sequence can exclude

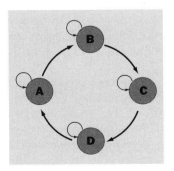

Fig. 4.6 The hypercycle. Each of the units A,B,C and D is a replicator. The rate of replication of each unit is an increasing function of the concentration unit immediately preceeding it. Thus the rate of replication of B is an increasing function of the concentration of A, and so on round the cycle.

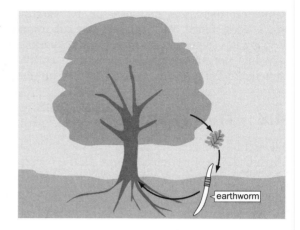

Fig. 4.7 An ecological hyper-
cycle.

another. The aid given by any molecule to the next one is ultimately fed back to
itself through the closure of the cycle. (The original model assumed translation
and coded protein replicases. This is not necessary, however: the members could
be ribozymes with replicase activity.)

For biologists, hypercycles may seem less mysterious if it is realized that eco-
systems are full of them. Figure 4.7 shows a two-member example: oak trees and
earthworms are replicators, and the presence of each accelerates the growth of
the other. Although ecologists usually speak of food chains, it is important to
remember that dead organisms ultimately provide nutrients for plant growth.
(English readers will be familiar with the cyclical nature of ecosystems from the
song 'On Ilkla Moor Baht 'at'.)

The hypercycle works fine until mutations are taken into account. There are
two types of mutation to consider (Maynard Smith, 1979a):

- mutations that alter an RNA so that it becomes a better or worse target for
 the replicase (Fig. 4.8a), and

- mutations that code for a better or worse replicase (Fig. 4.8b).

The fates of such mutations, through natural selection, are widely different, as
illustrated in Fig. 4.8. The failure of mutations of type 2 to spread means that
natural selection will not lead to the evolution of more and more efficient hy-
percycles.

Other difficulties have been pointed out for the persistence of hypercycles. It is
not always true that target efficiency will be optimized (Szathmáry & Demeter,
1987). A hypercycle can be destroyed by a 'parasite'—that is, a mutant that is
an excellent target, but a poor replicase (Bresch *et al.*, 1980). A large cycle may
be destroyed by a short-cut (Niesert *et al.*, 1981). Finally, a cycle with more than
four members will have widely fluctuating concentrations of its members: this

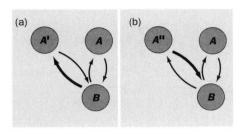

Fig. 4.8 Mutants in hypercycles. Initially, there is a two-unit hypercycle, $A \rightleftharpoons B$: (a) shows a 'selfish' mutant, A', that is a better target for B; (b) shows an 'altruistic' mutant, A'', that is a better replicator of B. Mutants of type A' will spread by natural selection in a homogeneous solution, but mutants of type A'' will spread only if the systems are isolated in compartments.

last objection echoes the argument (Pimm & Lawton, 1977) that ecological food chains are short because long chains are dynamically unstable.

It is worth remembering that all these arguments apply equally to ecosystems. Yet complex ecosystems do exist, and may owe their persistence in part to hypercyclic interactions of the kind illustrated in Fig. 4.7. It seems plausible that, because of hypercyclic interactions, and because of the self-inhibition of growth discussed in the last section, a diverse population of replicators could be stable.

Note that the hypercycle is not an individual in the sense that a bacterium is. It is, rather, a population of molecules interacting ecologically. Lacking individuality, it cannot be a unit of evolution (Szathmáry, 1989b). Molecules that 'sacrifice' themselves by producing replicases that serve the good of the hypercycle are 'altruists' in a sociobiological sense. To stabilize the hypercycle, one needs conditions in which altruists can spread, or at least coexist with 'cheaters'.

One obvious solution (Eigen *et al.*, 1981) is to put the hypercycles into compartments: that is, enclose them within a membrane. Then even improved replicase mutations can be selected for (Fig. 4.9). Imagine that, in a compartment, a favourable replicase mutant arises. When the compartment divides, the mutant will pass to one of the offspring. If the rate of compartment division is proportional to the growth rate of the hypercyclic population, the compartment with the mutant RNA will leave more descendents than those not having it.

To put hypercycles in compartments is, in effect, to create individuals with vertical transmission of genetic information, from parent to offspring. Given vertical transmission, the evolution of cooperation between the parts of an individual is to be expected. But if, as we have argued, early chemical reactions occurred on surfaces, we can ask whether cooperation could evolve without compartments, because neighbouring molecules would be genetically related. The same question arises, of course, in sociobiology: can kin selection give rise to a superorganism? The question is difficult. It is worth thinking about in the context of hypercycles, because of the relative simplicity of the interactions involved.

A start has been made by Boerlijst & Hogeweg (1991), who simulated spontaneous replication and decay, catalytic replication, and diffusion. As in Turing-type morphogenetic systems (see section 14.2), they observed spatial

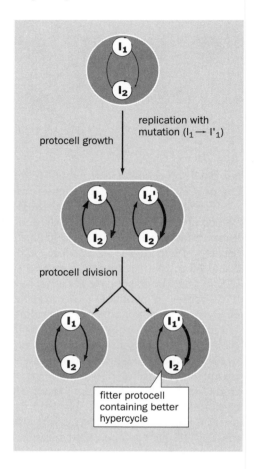

Fig. 4.9 Selecting improved replicase mutations by compartmentation of hypercycles. Molecule I′, arises by mutation from I, and is a better replicase.

patterns, in particular spiral waves. Within the spiral waves, members of the hypercycle predominate, and parasites (molecules that deny support to the next member of the cycle, but receive stronger than usual support from the preceding member) cannot spread.

These simulations suggest that cooperation between molecules may evolve more readily on a surface than free in solution. In any case, compartmentation with vertical transmission of genetic information is a satisfactory solution, but only if there are rather few RNAs in each compartment. If there are many such molecules present, selection favouring selfish mutants within compartments will predominate. The clothed, as opposed to the naked, hypercycle can not only benefit from favourable mutations: it can also protect itself from harmful ones—parasites and alternative cycles. The compartmentalized hypercycle is an efficient information integrator.

In the next section, we raise the following question: compartmentalized hypercycles seem to be OK, but once we have compartments, do we need a hypercycle inside?

4.6 The stochastic corrector model

Sociobiologists have known for some time that population structure facilitates the survival of altruists. The prebiological application of this idea is due to Michod (1983). He starts with a model similar to Wilson's (1980) 'group selection' model (Fig. 4.10). RNA molecules, initially in a well-stirred homogeneous solution, become absorbed randomly to small sites on a surface. They undergo a few rounds of replication there (Wilson's original model did not allow reproduction in his groups). There are two types of RNA: one encodes a replicase at a cost in replication speed to itself, and the other is a parasite—it is replicated but does not perform any function useful to the system. Obviously, in a homogeneous solution, the replicase-coding RNA would go extinct, followed by the parasite. Not so, however, on the surface with population structure: the small groups on the rock are assembled randomly, and contain varying proportions of the two types of RNA. Replication will be faster in those groups with a higher

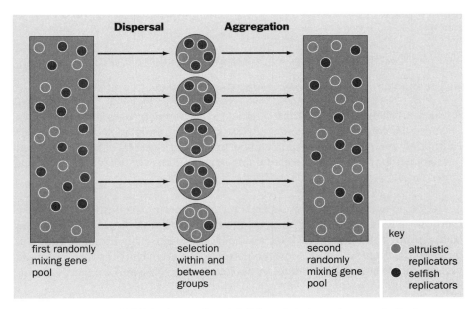

Fig. 4.10 Wilson's (1980) 'trait group' model. If there is to be an increase in the frequency of altruistic replicators, as shown, then either (1) altruistic replicators must assort together non-randomly or (2) there must be synergistic fitness effects, so that a group of altruistic replicators contributes disproportionately to the next generation.

proportion of useful molecules, although, within each group, the selfish molecules will replicate faster. Imagine that local populations are regularly washed away by the tide from the rock surface and mixed thoroughly, and that new groups are then formed as small random samples of these molecules. The majority of the offspring RNAs will have come from groups that had a high proportion of replicase genes. There will be selection between groups favouring the replicase genes, and within groups favouring selfish genes. It can be shown that, provided new groups consist of very few molecules, there will be a stable gene/parasite polymorphism at the level of the whole population.

Population structure is most pronounced when we have replicatable compartments, each containing only a few molecules: the variance between groups is then at its highest. Michod (1983) wrote 'Thus, initially, the organism was one extreme of the population structure of the macromolecules of life. The organism was the culmination of the encapsulation phase of evolution which proceeded through initial phases of passive localization'. In other words, the first stages of cooperation between replicators may have originated simply because they were neighbours on a surface: only later were discrete compartments, or protocells, formed, within which cooperation could evolve further.

We now turn to an explicit model, the 'stochastic corrector' model (Szathmáry, 1986; Szathmáry & Demeter, 1987), in which compartments contain reproducing groups of genes (Fig. 4.11). The assumptions of the model are as follows:

Two kinds of replicating molecules are contained within compartments.

- The number of molecules per compartment is small.

- The rate of division of a compartment depends on the types and proportions of molecules contained. There are 'altruistic' and 'parasitic' molecules: the latter multiply more rapidly within a compartment, but the total rate of replication is greater for compartments containing equal numbers of the two types, and is zero if one type of molecule is absent.

The behaviour of the model depends on two types of stochasticity: (1) Since there are few replicators per compartment, the outcome, from a given initial composition, is not determined. Occasionally, the intrinsically less-fit variant will outgrow the fitter. The only constant feature is the probability distribution of the various compositions. (2) At division, molecules segregate into compartments at random.

Computer analyses showed that beneficial but 'altruistic' mutants are able to spread: compartments harbouring copies of such mutants divide more frequently. Also, parasites are successfully selected against at the compartment level. Another important feature is that there is selection favouring compart-

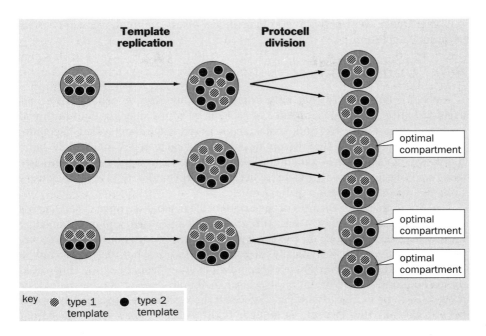

Fig. 4.11 The stochastic corrector model. The initial compartments contain three copies of type 1 templates and three copies of type 2 templates. During template replication there is a tendency for 'black' templates to outgrow 'whites', but this does not proceed deterministically, due to the small number of templates within each compartment. Templates are randomly distributed at protocell division. The stochastic nature of replication and redistribution ensures that the optimal type of compartment reappears in the next generation. The population settles down to an equilibrium with a constant proportion of optimal compartments.

ments with synchronized gene replication, because such synchrony helps to ensure that the composition of daughter cells will resemble that of the mother cell.

All in all, the stochastic corrector model is an efficient information integrator, and as such it is a satisfactory resolution of Eigen's paradox. But we have seen that hypercycles within compartments are also satisfactory. The relationship between the models is as follows. A hypercycle is a set of replicators that, despite the objections listed above, could have a degree of persistence in the absence of localization or compartmentation. Once compartments exist, however, the stochastic corrector model shows how a much wider range of possible compositions can be replicated. For example, the genes in a single compartment need not all have replicase activity, as is the case for the hypercycle. It would be sufficient that one be a replicase: the others could code for enzymes useful for the system as

a whole. Hence the stochastic corrector model opens up greater possibilities for continued evolution.

4.7 Conclusions

A replicator is an entity that only arises by the division or copying of a pre-existing entity. The mere existence of replicators is not sufficient for continued evolution by natural selection, which requires what we have called 'indefinite hereditary replicators': that is, entities that can exist in an indefinitely large number of states, each of which can be replicated. In living organisms, nucleic acid molecules are the only indefinite hereditary replicators, or at least they were until the invention of language and music.

A major problem for theories of the origin of life is what we have called Eigen's paradox. The amount of information that can be transmitted, and maintained by natural selection, depends on the accuracy of replication. In existing organisms, accuracy is achieved by genetically programmed enzymes, but these could not be primitive. Without such enzymes replication accuracy would be low. How could the total quantity of information increase? In the absence of compartments, or protocells, some increase could be achieved if the replicators interacted to form a hypercycle, but this increase is limited, both because hypercycles can be destroyed by parasites, and because large hypercycles are dynamically unstable.

Some further increase in information may be possible if interactions are confined to a surface. The decisive step, however, took place when replicating molecules were confined within compartments. If the number of molecules per compartment is small, stochastic effects will generate variability upon which selection can act. If the rate of growth and division of the compartments depended on the kinds of molecules they contained, division of labour and co-operation between molecules would evolve.

5 The Chicken and Egg Problem

5.1 Introduction

The most fundamental distinction in biology is between nucleic acids, with their role as carriers of information, and proteins, which generate the phenotype. In existing organisms, nucleic acids and proteins mutually presume one another. The former, owing to their template activity, store the heritable information: the latter, by enzymatic activity, read and express this information. It seems that neither can function without the other. Which came first, nucleic acids or proteins? There are three possible answers: (1) nucleic acids; (2) proteins; (3) neither: they coevolved.

In this chapter, we discuss various possible answers to this 'chicken or egg?' problem. In section 5.2, we discuss what seems to us the most likely answer, that at first RNA performed both functions, as replicator and enzyme. In section 5.3, we consider an alternative view, in which protein enzymes existed either before, or alongside, the first nucleic acids. In section 5.4, we ask whether, perhaps, the first replicators were not nucleic acids. Finally, in section 5.5, we ask why, given that the genetic message is carried by nucleic acids, there are only four nucleotides and two base pairs.

5.2 RNA as an enzyme

So far, we have tacitly assumed nucleic acids preceeded proteins, without stating the main reason. Nucleic acids came first because they can perform both functions: they are replicable, and they can have enzymatic activity. For many years, a common opinion was that to be replicable almost amounted to self-replicative ability, but that it was far-fetched to assume enzymatic activity. Today, there is increasing evidence that RNA can act as an enzyme, but we are more aware of the difficulty of self-replication.

It should have been expected on theoretical grounds that RNA could act as an enzyme: the possibility was discussed by Woese (1967), Crick (1968) and Orgel (1968). Consider first why proteins can act as enzymes. An enzyme has a well-determined three-dimensional structure of chemical groups that, in most cases, arises automatically from the primary structure. Substrates of the enzyme are bound by the chemical groups on the surface. This means that the reactants will be kept in close proximity, and hence experience a much higher local concentration of each other than in solution. This by itself increases the rate of the reaction.

Enzymes speed up reactions in a second way. During a chemical reaction, an intermediate structure of high energy is formed, the so-called transition state complex. The higher this activation energy, the slower the reaction. As Linus Pauling suggested 50 years ago, enzymes decrease the activation energy by binding to and distorting the reactants. Further, enzymes are not rigid

structures: by torsion and bending, they may guide the reaction. The effect of these various processes can be to increase the reaction rate by more than a millionfold. After the reaction is complete, the products leave the enzyme, and the latter, having completed its catalytic cycle, is ready to start the next one.

What of nucleic acids? RNA often forms well-defined, flexible, three-dimensional structures, presenting various functional groups on its surface. In principle, therefore, one would expect some RNA molecules to act as enzymes. In the first clear demonstration of such action, Kruger *et al.* (1982) showed that the splicing of a pre-rRNA transcript in the macronucleus of the protozoan *Tetrahymena* can proceed without help: the intron splices itself out from the molecule, and the two exons are joined together (Fig. 5.1). In this example, however, the biochemical device—the intron—does not complete a full cycle: it does not reappear in its original form at the end of the process. Therefore we cannot call it a catalyst *sensu stricto*.

Enzymatic activity proper was soon discovered by Zaug & Cech (1986). The final product of the splicing process just described is an intervening sequence, IVS, 19 nucleotides shorter than the original intron. This IVS RNA acts as a proper enzyme in the following reactions with cytidine:

$$IVS + C_5 \rightarrow (IVS - C) + C_4,$$
$$(IVS - C) + C_5 \rightarrow IVS + C_6$$

$$\overline{C_5 + C_5 = C_4 + C_6.}$$

The 19-nucleotide IVS drops out of the overall equation: hence, the reaction is truly catalytic. The catalysed reaction is

$$C_i + C_j = C_{i+1} + C_{j-1},$$

where $i >= 3$ and $j >= 4$. The end result of the reaction is a mixture of poly(C) molecules of different lengths, with a maximum of 30 nucleotides. Note that no energy input is necessary for these transesterification reactions: hence, the mixture is at equilibrium.

This reaction is important not only because the RNA acts as a real enzyme (ribozyme), but also because there is a net elongation of certain substrate RNAs. Later it was found that another variant of the IVS can attach C and U to C5, producing molecules up to 20 monomers long.

After this it was natural to try to make an RNA that could act as a true polymerase, copying an external template. Doudna & Szostak (1989) managed to modify the *Tetrahymena* IVS so as to enable it to catalyse complementary strand synthesis of an external template. This ribozyme, however, is not very efficient. The maximum length of template that can be copied is limited to 40, and only 1 per cent of such templates are copied completely. Doudna *et al.* (1991) constructed a ribozyme of three subunits, by cutting the original self-

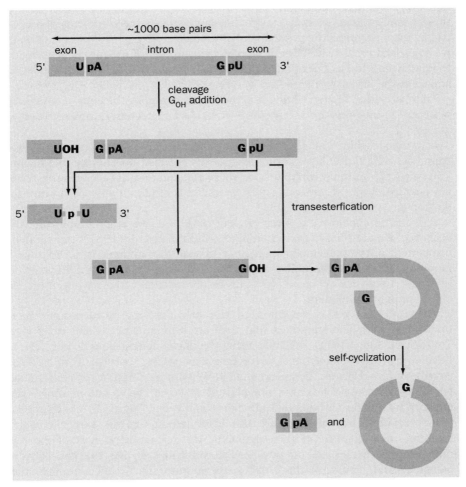

Fig. 5.1 Self-splicing of *Tetrahymena* pre-rRNA (after Darnell *et al.*, 1986). The hydroxyl group of the guanosine (G_{OH}) attacks the U–pA bond at the 5′ exon–intron junction, and is added to the intron. The intron then undergoes further self-catalysed transformations.

splicing intron, *sunY*, from bacteriophage T4 into three. The logic behind this technique is that, while the subunits can assemble together by secondary bonds to form an active enzyme, the subunits themselves are kept to copiable size. This enzyme turned out to be able to copy one of its subunits by catalysing the covalent linking of pre-formed oligonucleotides on the template. This is still far short of a true self-replicating ribozyme for three reasons: (1) only one of the three subunits is copied; (2) the building blocks are not monomers; and (3)

template recycling, including $+$ and $-$ strands, would be necessary. Despite these complications, we may expect that quite soon a truly autocatalytic RNA will be made, opening the possibility of autonomous RNA coevolution *in vitro*, as we suggested previously. From the prebiological point of view it is encouraging that activated nucleotides can condense into chains 20 to 40 nucleotides in length in the absence of templates (Joyce, 1989).

There are other known ribozyme activities: for example, the active component of RNAse P, that cuts the 5′ end of pre-tRNAs, is a ribozyme (Guerrier-Takada *et al.*, 1983).

It is remarkable that all these enzymatic activities relate to nucleic acid substrates, and utilize base pairing as a means of substrate–enzyme interaction. This raises the question whether ribozymes acting on non-nucleic acid substrates are possible at all. Needless to say, the answer is of paramount importance for our topic.

There is no conclusive answer yet, but there are two reports indicating the existence of such ribozymes. According to Shvedova *et al.* (1987), the catalytic component of the polyglucan branching enzyme in mammals is an RNA with many modified bases. Recently, Yanagawa *et al.* (1990) found that, in yeast, 5-hydroxycytidine can act as catalyst with oxidoreduction activity without the aid of any protein component.

White (1976) called attention to the fact that many coenzymes have nucleotide parts, and suggested that they are remnants of ancient ribozymes. This belief is supported by the finding that the nucleotide moieties do not seem to have an essential role in the catalytic reactions proper: their role is to bind the enzyme to its substrates. Benner *et al.* (1989) point out that, in some enzymatic reactions, nucleotide cofactors are present in some cases, but in others are replaced by simpler variants that function equally well (Fig. 5.2). They suggest, therefore, that the cofactors are the fossil remains of an earlier complex ribozymic metabolism. On this argument, the reason why most of the few remaining ribozymes act on nucleic acid substrates is that the possibility of complementary base pairing makes them particularly effective for such reactions. At an earlier time, before the evolution of translation, ribozymes were involved in a much wider range of reactions: the remaining nucleotide cofactors are evidence of this, and provide clues to the nature of primitive ribo-organisms.

A few years ago, Szathmáry (1984, 1989b, 1990) suggested a systematic way of testing the enzymatic capabilities of RNAs. Let us choose the reaction that we want to be catalysed. We must know the structure of the transition-state complex of the reactions, and we must be able to synthesize a close, stable analogue of it. It is known that antibodies produced against such analogues can catalyse the original reaction (Tramontano *et al.*, 1986; Pollack *et al.*, 1986). *Mutatis mutandis*, if we manage to produce RNAs that tightly bind to these analogues, we may find that they, too, can act as ribozymes.

Fig. 5.2 Cofactors with and without RNA fragments. The presence of the simpler variants suggests that the RNA fragments are a historical relict, not performing any essential function (after Benner *et al.*, 1987).

But how can we produce an RNA, the structure of which is not known in advance? The answer is that we do not do it: let natural selection take care of the job. Note that during the formation of catalytic antibodies the strategy is also evolutionary: the immune system produces them through variation and selective amplification.

We know from Spiegelman's experiments that RNA phenotypes can be selected. Orgel (1979) suggested that the *in vitro* selection of RNAs binding small molecules could be possible. In our case, these small molecules are the transition-state analogues. The selection step can be carried out by affinity chromatography, using the analogue as the target. Independent replication and selection

Fig. 5.3 Experimental design for selecting RNA. The procedure starts with a randomized DNA library of sequences. The selection–amplification–selection cycle is repeated until strongly binding RNAs are obtained. The reason for amplifying at the DNA level is that there is no competition between different sequences, as there would be if RNA was amplified.

were predicted to lead to tightly bound RNA-small molecule complexes, and possibly ribozymes (Szathmáry, 1989b, 1990).

This ambitious programme has already been partly realized in two laboratories. Tuerk & Gold (1990) wanted to select for RNAs that can tightly bind to the protein DNA polymerase of phage T4 (Fig. 5.3). Normally, a certain region of the mRNA for this protein binds to the protein, contributing to translational control. Tuerk & Gold chose to vary the eight nucleotides in the loop of the RNA binding site. They synthesized all the 65 536 possible RNAs. Selection was carried out on a sieve containing the protein, and amplification was done at the DNA level by the polymerase chain reaction (PCR) to ensure higher fidelity and no competition among the oligomers (Fig. 5.3). Two best-binding RNAs were selected: one was identical with the wild type, another was four mutational steps away from it.

Ellington & Szostak (1990) wanted to select for RNAs binding small molecules such as Reactive Blue and Cibacron Blue. They synthesized 10^{13} different random RNA sequences that were about 100 nucleotides long. This is more like the biological situation in genetics, where the number of existing individuals is far smaller than that of possible sequences. RNAs were selected by affinity chromatography; one of them is shown in Fig. 5.4. In this way, they obtained tightly binding RNAs that were specific: cross-binding to related molecules was weak. Also, several different solutions to the binding site problem were found. It

Fig. 5.4 An RNA molecule selected to bind to a specific substrate.

is estimated that for every small molecule, there are some 100–1000 radically different sequences, among the possible 10^{60}, that can bind rather well to the ligand. If, however, all the functional mutants of these sequences are also taken into account, the estimated number of binding sequences amounts to an impressive 10^{49}–10^{52}.

We are happy to see that Prudent *et al.* (1994) apparently have made a breakthrough. Following the described protocol above, they were able to select, using a transition state analogue, a ribozyme that catalyses the isomerization of a bridged biphenyl.

5.3 Autocatalytic protein nets

Fox (1984) has strongly advocated a primary role for proteins in prebiotic evolution. His thermally formed proteinoids have the interesting feature of spontaneous weak catalytic activity. Catalysed reactions include esterolysis, phosphatolysis, decarboxylation, deamination, oxidation and photochemical decarboxylation. The composition and sequences of amino acids are not completely random because, during polymerization, the next amino acid to be added to a chain depends on the amino acids already present.

It is important that catalytic activity should have been demonstrated, but proteinoids do not lead us very far unless they are capable of evolution by natural

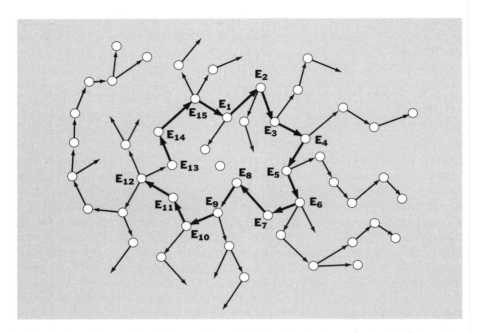

Fig. 5.5 A reflexively autocatalytic protein network (after Eigen, 1971). Participating proteins (E_i) are assumed to catalyse several different reactions (cleavage and condensation). In order to form an autocatalytic set, the formation (from shorter oligopeptides) of each member of the set must be catalysed by other members of the same set. Proteins E_1–E_{15} constitute such a set: other proteins in the network are not part of the cycle, and are in effect waste products.

selection. It seems, however, that the first necessary requirement is lacking: they do not replicate. They can catalyse the formation of very short oligonucleotides from free amino acids, but this is very far from self-reproduction.

Imagine, however, that some proteinoids are able to catalyse the further joining of peptides. Eigen (1971) suggested that, in that case, a catalytic cycle might be possible, in which every enzyme catalyses the formation, from simple peptides, of the next one, without template replication of the cycle members (Fig. 5.5). Thus, there would be no autocatalytic member of the cycle, but *the system as a whole increases autocatalytically*. Such a cycle is very different from the autocatalytic cycles we discussed in section 3.4. Such cycles produce two copies of a molecule *A* from a single *A*, but the individual steps in the cycle are not themselves catalysed. In contrast, we are now imagining a system in which all reactions are catalysed, and in which the catalysts themselves are products of the system.

It is important to emphasize that no such system exists today: known examples are very remote from it. For example, in the synthesis of the antibiotic gramicidin S, which is made of 10 amino acids without translation from a

nucleic acid sequence, the necessary information is carried by the enzymes themselves. Note that in this case five large enzymes are needed to synthesize an oligopeptide with a specific amino acid sequence.

The following questions arise:

- Is such an autocatalytic cycle possible at all on probabilistic grounds?
- If so, can it arise spontaneously from a disordered state?
- If so, is there a possibility of heredity?

We will consider these question in turn.

Kauffman (1986) argued that the answer to the first question is yes. He reasons as follows. Suppose that peptides are formed from free amino acids. Although this condensation liberates water, peptides of sufficiently large size can appear spontaneously at high amino acid concentrations. The so-called plastein reaction serves as a crude example. In this reaction, the digestive enzymes pepsin and trypsin, as well as cutting proteins, also catalyse transpeptidation reactions: from a concentrated solution of oligopeptides, they catalyse the formation of longer peptides, until an equilibrium mixture is reached. If some of the longer peptides are removed, the equilibrium is restored by their reformation. Note that the reaction does not require an input of energy. This is analogous to the formation of longer oligonucleotides from shorter ones, which, as we have seen, does not need an energy input either.

For simplicity, Kauffman supposed that peptides are formed from only two kinds of amino acids, A and B, by simple cleavage and ligation reactions, such as $AAB + BB \rightarrow AABBB$. He assumes a constant probability, P, that any peptide catalyses any reaction. Hence a single reaction may be catalysed by several enzymes, and the same enzyme may catalyse several reactions. Knowing the value of P, what is the probability that a network consisting of peptides with a maximum length of M will contain an autocatalytic set? Kauffman calculated that this probability is high, even for small P, provided that exchange reactions such as $AABA + BBBB \rightarrow AABB + A + BBB$ are allowed. Suppose, for example, that the maximum length of a peptide, M, is 13: there are then some $18\,000$ possible peptides. Even with P as low as 10^{-9}, an autocatalytic set exists with probability 0.999.

The basic model neglects several complications. What if there is a threshold size M, below which a peptide is incapable of catalysis, as is likely to be the case? The calculations show that even if this threshold size is 50, reasonably sized autocatalytic sets still appear with good probability. The calculation also assumes that a peptide of length, say, 100 can recognize without mistakes a substrate peptide built of, say, 200 amino acids. This is obviously nonsense. Kauffman recognized the problem, and pointed out that in reality an enzyme is likely to recognize some basic common features in many different peptides, and to react with all of them. In consequence, many extraneous reactions will occur, but this does not alter the conclusion that reflexively autocatalytic subsets will exist.

Orgel (1987) raised a more serious objection to the scheme. If, for example, the maximum length of a peptide is 13 amino acids, there are then some 18 000 possible peptides of length 13 or less, but over 100 000 different bonds. It is not plausible that such a short peptide should catalyse 5–10 different specific reactions.

It is not clear, therefore, whether autocatalytic sets are chemically possible. Even if they are possible, are they simple enough to have a chance of coming into existence? Logically, if all peptides below the critical size are present, the cycle starts running right away. But it is too generous to assume such an initial condition, particularly if the critical size is large. It is more likely that some necessary members of the autocatalytic set will be absent. Is it possible for the mixture to jump from chaos to order? A 'toy' model by Dyson (1985) indicates that the answer is yes: there is a possibility of a transition from chaotic to organized complexity.

Dyson imagines a population of droplets, or compartments, containing amino acids and peptides. Initially, the droplets neither divide nor die: there is no Darwinian selection. A unit step is the replacement of one amino acid in a peptide by another. Each amino acid is either 'active' or 'inactive'. The definition of an active monomer is circular: a peptide containing active monomers is more likely to cause the incorporation of active monomers in other peptides. Note that, by this definition, Dyson is assuming that an autocatlytic set of peptides is chemically plausible—the point that Kaufmann was attempting to establish by a probability argument.

Suppose that, at any time, the proportion of active monomers is x, and that $y = F(x)$ is the probability of insertion of an active monomer. Dyson argues, on kinetic grounds, that the graph of y against x will be an S-shaped curve, cutting the $x = y$ line at three points (Fig. 5.6). These points of intersection are equilibria. If $y > x$, then x will increase, and if $y < x$ then x will decrease. Hence, points A and C are stable, and are separated by an unstable equilibrium at B. Dyson calls states A and C 'dead' and 'alive', respectively.

Can the system jump spontaneously from A to C? If the number of amino acids in a droplet, N, is large, the system will behave deterministically, and a jump from A to C is not possible. The transition is possible only if stochastic events are important, and hence only if N is not too large. Using plausible parameters, Dyson estimates that for $2000 < N < 20\,000$ a jump is likely.

A remarkable feature of the model is its error tolerance. After a jump to C, the probability of incorporation of a correct amino acid can still be as low as 0.75. Such a low value would be quite inadequate for a system depending on template reproduction, but is acceptable here: indeed, the origin of an autocatalytic set requires this high tolerance.

Given that a jump from A to C is possible, can the reverse jump also happen? The answer is that such reverse jumps will happen, but need not destroy the system. Thus, once C is reached, the system will grow autocatalytically, and,

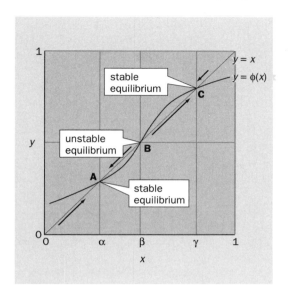

Fig. 5.6 The dynamics of Dyson's (1985) model. x is the proportion of monomers incorporated into peptides that are 'active', in the sense of favouring the incorporation of other active monomers into peptides; $y = \phi(x)$ is the probability that a new monomer inserted into a peptide will be active. When $x = \phi(x)$, the value of x is at an equilibrium. The upper (C) and lower (A) equilibria are stable, and the central one is unstable. Transition from the lower 'dead' state to the upper 'alive' state depends on stochastic events.

through compartment fission, more and more C compartments will be produced. Hence we can accept that a jump from chaos to order may be feasible. There is, however, some difficulty in reconciling Dyson's model with Kauffman's. Note that Dyson estimated 20 000 amino acids as an upper limit if a jump is to occur: Kauffman estimated approximately 20 000 peptides (*not* amino acids) as the number required to make an autocatalytic set likely. Given the over-simplified nature of the models, however, this discrepancy need not be fatal. In any case, there was plenty of time for rather unlikely transitions to occur, so Dyson's estimate of $N < 20\,000$ may be needlessly severe.

Even if autocatalytic protein networks are chemically possible, and are simple enough to spring to 'life' as well, there remains the problem of their evolutionary potential. It was the apparent lack of heredity that made Eigen (1971) abandon his own idea. Figure 5.7 illustrates the problem: there is no guarantee that a 'mutant' enzyme E_i' would initiate a complete mutant cycle, which ultimately would have to feed back onto E_1'.

It is plausible, provided that autocatalytic sets are possible at all, that alternatives can occur within a large protein network. These alternatives would, presumably, be dynamically stable. Hence there would be heredity within, and selection among, the alternative sets. Indefinite evolution would, however, be ruled out because of the limited range of hereditary variation. The unique feature provided by nucleic acids is the indefinitely large number of chemically stable, replicable, close alternative sequences that are possible.

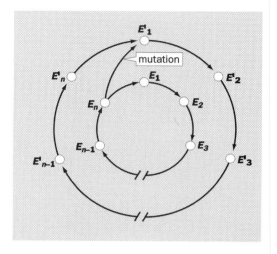

Fig. 5.7 A mutant in an autocatalytic cycle. The inner cycle is the ancestral one. In a typical 'mutation', E_n gives rise to E_1' instead of E_1. For heredity, and hence selection, to operate, it is necessary that E_1' should initiate a new cycle that ultimately feeds back on itself, as shown in the outer cycle: unfortunately, there is no guarantee that this will happen.

If, however, autocatalytic sets of proteins of the kind envisioned by Kauffman and Dyson existed, they would have generated a large number of chemical compounds that were irrelevant to the maintenance of the system. Nucleic acids could have been among these compounds. Initially they would have been parasitic on the autocatalytic system, but if there was compartmentation, the compartments in which RNA developed some functionality would have been better off. Finally, RNAs would turn from parasites into symbionts, and act as genetic material (Dyson, 1985).

5.4 The urgene: RNA, clay or something else?

So far, we have tacitly assumed that the first genetic material was RNA. The selection experiments encourage us to think that we have found the chemistry that can link the worlds of chemistry and biology. Granted that this view is replication-chauvinist, we must still ask whether chemical evolution could have produced replicative RNAs. The beautiful theoretical work on RNA selection almost makes us forget that the spontaneous generation of RNA is highly problematic. As we saw in Chapter 3, even the formation of nucleotides is problematic: pyrimidine-containing ones could not be synthesized.

Let us for the moment forget these difficulties, and assume that nucleotides are available. A major snag is that chemically synthesized nucleotides are expected to consist of an array of isomers. Joyce *et al.* (1987) pointed out that isomeric impurity can easily block replication. Ribose exists in two stereoisomers, differing only in the orientation of one –OH radical. Only D-ribose is used to make RNA, but both arise in equal concentrations in prebiotic experiments. A further

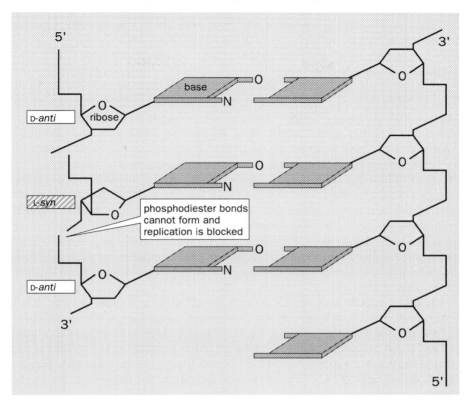

Fig. 5.8 Disruption of RNA replication by isomeric impurities (after Joyce *et al.*, 1987). Only D-*anti* isomers are used in biological systems, as shown on the right. An L-*syn* isomer (differing in the orientation of an –OH radical and in the relative orientation of ribose and base in the nucleoside) can pair with the complementary D-*anti* nucleotide in the template, as shown on the left. The formation of a phosphodiester bond is then impossible, and replication is blocked.

complication arises with the relative orientation of ribose and base in nucleosides: again we have two isomers, *syn* and *anti*. Biological nucleotides have the D-*anti* conformation. Imagine that a fully correct RNA template is to be copied in a mixture consisting of activated D-*anti* and L-*syn* nucleotides. As shown in Fig. 5.8, replication is blocked.

Such problems have led Cairns-Smith (1971) to suggest that primordial genes were made, not of RNA, but of clay. This biblical idea is not entirely unreasonable. Clay crystals are readily formed from a saturated solution of the necessary ions. They grow around crystal nuclei, and, once grown, may fall into several pieces, each piece being able to grow again. But can such a system undergo evolution?

It is clear that clay crystals can multiply, but the question of heredity is harder to answer. A crystal, through errors in the lattice, can store information. If, and only if, the pattern of errors is replicated can we speak of heredity. It is also possible to imagine circumstances in which different crystals could be differently selected. Imagine a porous sandstone into which the saturated solution is leaking. From this, clay begins to crystallize. Crystals with different errors might grow at different speeds. Crystals that are too loose might be easily washed away: those that are too dense might block further percolation of the saturated solution. It is therefore conceivable that clay crystals could have hereditary variation that affected fitness. If so, they could be units of evolution.

If for the moment we accept this conclusion, why and how were clay genes replaced by RNA genes? The why is easy to answer with hindsight: molecular information storage and retrieval is far more efficient. The how can be imagined only gradually, in small steps. Some organic compounds can be bound by clay minerals, and some of them may enhance crystal growth. If clay genes had been able to manipulate their organic environment, then sooner or later the synthesis of RNA could also have become feasible. At first, RNA may have acted to stabilize the clay.

The main attraction of the idea lies in the difficulty in explaining how RNA could have arisen spontaneously. In Cairns-Smith's analogy, if we see an arch, we are tempted to think that someone constructed a scaffold first, used it to build the arch, and then removed the scaffold. The arch is the RNA: the scaffold is the clay. However, a good scaffold should be such that it can be constructed on its own, and should be able to carry the weight of the stones of the arch. Unfortunately, there is no strong evidence that clay minerals satisfy either criterion. Heredity has not been convincingly demonstrated. Even if clay minerals are able to evolve, it is not guaranteed that they could have reached the complexity of an RNA 'breeder'. It seems better, therefore, to seek more conventional ideas about what the scaffold may have been made of.

Schwartz & Orgel (1985) suggested that ribose was replaced in the urgene by something simpler—for example, glycerol phosphate. Other possibilities were suggested by Joyce *et al.* (1987). Such analogues (Fig. 5.9) would not have suffered from the severe problem of enantiomeric cross-inhibition described

Fig. 5.9 Some possible alternatives to ribose.

above, owing to the free rotation of the glycerol–base bond, but would have retained the essential feature of complementary base pairing. Wächtershäuser (1992), however, pointed out that these intermediates are not compatible with his notion of a prebiotic pizza. He suggested instead 'tribonucleic' acids, in which the backbone of the double-stranded molecule is formed by covalently linked phospho-trioses, which could readily have formed on the surface. Further, to cope with the non-formation of pyrimidine nucleotides, he suggested an all-purine version of nucleic acids, where the non-standard nucleosides are bonded via their N in position 3, instead of the normal N9 bond. The question, however, is still open: As yet, no-one has produced a prebiotically feasible replicating RNA analogue.

To sum up, we do not know how the primordial genetic material was formed. However, the hypotheses discussed have one thing in common: all assume that the urgene was formed without a significant participation of protein.

5.5 What determines the size of the genetic alphabet?

It is a commonplace that the genetic alphabet consists of four letters, if we take U and T as essentially equivalent. Why should this be so? It cannot be argued that there are only two possible base pairs, because Piccirilli *et al.* (1990) have shown that it is possible to design, and even to synthesize, novel pairs (Fig. 5.10). Isocytosine and isoguanine have been known for some time. They could not, in fact, be used as a novel base pair, because their H-bonding patterns are not stable enough: they rearrange through tautomerization rather frequently. Thus, they are often used as mutagenic agents. This is not true for the other novel pairs, however. Piccirilli *et al.* (1992) were able to synthesize the κ–χ pair by organic chemical means. Both RNA and DNA polymerase accept and copy this new pair.

What difference does it make whether the genetic alphabet is small or large? Orgel (1990) suggested that evolution either (1) failed to 'discover' more base pairs, or (2) 'decided' that two pairs were enough. Option (1) is hard to test, but one can wonder what could have led to option (2). It might be the case that three pairs would be too many, and one pair too few. This implies that genetic alphabets made of two pairs could be optimal in evolutionary terms; that is, organisms with two pairs would have had the highest fitness.

Szathmáry (1991b) suggested a theory for such an evolutionarily optimal state. The underlying idea is very simple. One has to go back to the RNA world, when RNAs served as genetic material as well as ribozymes. It is intuitively clear that the more monomer types we have, the higher the catalytic efficiency of macromolecules can be. Metabolic efficiency of ribo-organisms, and hence their growth rate, would increase with increasing size of the genetic alphabet. This

Fig. 5.10 Novel base pairs (after Piccirilli *et al.*, 1990).

by itself does not lead to an optimum; some other effect must counterbalance the advantage gained from more letters. This other effect results from decreasing copying fidelity: an increase in the number of letters increases the mutation rate, and hence the mutational load.

The effect of alphabet size on mutation rate can be appreciated by the following simple example. Take a purine with an H-bonding pattern A–D–A, for example. Its complementary pair would have a D–A–D pattern:

$$
\begin{array}{ll}
\text{A–D–A} & \text{purine} \\
\text{| | |} & \\
\text{D–A–D} & \text{pyrimidine}
\end{array}
$$

If one wanted to choose another pair, then it would be reasonable to take a purine with D–A–D, and a complementary pyrimidine with A–D–A configuration:

$$
\begin{array}{ll}
\text{D–A–D} & \text{purine} \\
\text{| | |} & \\
\text{A–D–A} & \text{pyrimidine}
\end{array}
$$

It is easy to see that the likelihood of transitions is minimized by this choice. Thus the mismatched pairs

$$
\begin{array}{lll}
\text{A–D–A} & \text{D–A–D} & \text{purine} \\
* & * \quad * & \\
\text{A–D–A} & \text{D–A–D} & \text{pyrimidine}
\end{array}
$$

have no complementary A–D interaction at all; moreover, the ∗ indicates active repulsion caused by steric clash between two H-donors.

With two base pairs, therefore, replication accuracy can be high. But if a third base pair is added, some complementarity between mismatched pairs is inevitable. For example, a third purine A–D–D forms the following mismatches with the previous pyrimidines

$$
\begin{array}{ll}
\text{A–D–D} & \text{A–D–D} \\
\text{| | }* & *\text{ |} \\
\text{D–A–D} & \text{A–D–A}
\end{array}
$$

which necessarily leads to a dramatic decrease in fidelity, since now mismatched pairs are attracted to each other somewhat by partial complementarity. For this reason, Szathmáry found that the decrease in copying fidelity with alphabet size is faster than exponential.

The fitness costs of inaccurate replication depend on genome size: the larger the genome, the greater the fitness cost. Since we are still in an RNA world, it is reasonable to suppose that there was no mismatch repair or proofreading, and hence that the genome size was limited to 10^3–10^4 bases.

Given these assumptions, one can estimate the fitness costs of an increased alphabet size. It is much more difficult to obtain the relationship between alphabet size and metabolic efficiency. Szathmáry (1992) attempted such a calculation. If, as is approximately true, four monomers participate in an active site, the number of possible active sites is m^4, where m is the number of kinds of monomer. This gives 256 different active sites for ribozymes with four monomer types, and 1.6×10^5 for a protein enzyme made from 20 kinds of amino acid. These figures are underestimates, because of post-transcriptional modification, and the possible use of cofactors discussed above, but they do give a rough idea of the relative numbers of possible active sites, and hence of the relative accuracies with which such sites could evolve to fit particular substrates.

Although these figures give some idea of the relative efficiencies of different alphabet sizes, before we can estimate an optimum alphabet size we need to have some idea of absolute efficiencies: how big an alphabet is it really useful to have? We know that at least some protein enzymes are 'perfect', in the sense that the rate of the catalysed step equals that of diffusion of the reactants to and from the enzyme. However, the fact that translationally modified amino acids sometimes play an important role in increasing catalytic efficiency suggests that an alphabet of 20 amino acids is not sufficient to ensure that a perfect enzyme can be produced for every substrate. It follows that an alphabet of 20 approaches, but does not quite reach, the size needed to generate a fully efficient set of enzymes. The calculation is further complicated by the fact that one must consider, not a single reaction, but a metabolic network. Despite these difficulties, however, it turns out that a rather robust optimum exists, with an alphabet of four monomers, or two base pairs.

We think, therefore, that for ribo-organisms four bases were optimal. No such optimum argument applies to modern organisms with protein enzymes. The genetic alphabet is a frozen character state that evolved when enzymes were ribozymes. Once protein synthesis and the genetic code had evolved, the catalytic repertoire could be extended by the addition of novel amino acids, without lowering the accuracy of replication.

One final point: the theory of optimal genetic alphabet size is testable. Copying fidelity is not very difficult to measure, and, by application of RNA breeding *in vitro*, it would be possible to generate ribozymes catalysing the same reaction, but made of different numbers of letters, and to compare their efficiencies.

6 The Origin of Translation and the Genetic Code

The origin of the code is perhaps the most perplexing problem in evolutionary biology. The existing translational machinery is at the same time so complex, so universal, and so essential that it is hard to see how it could have come into existence, or how life could have existed without it. The discovery of ribozymes has made it easier to imagine an answer to the second of these questions, but the transformation of an 'RNA world' into one in which catalysis is performed by proteins, and nucleic acids specialize in the transmission of information, remains a formidable problem.

We start, in section 6.1, by discussing changes known to have occurred in the code since its origin. Although these changes are minor, they do shed some light on how the code may have evolved in its very early days. In section 6.2, we ask what can be deduced from the present assignment of codons to amino acids, and from the phylogeny of tRNAs. Finally, in section 6.3, we come to grips with the hardest question: how did a specific association between particular amino acids and particular codons first come into existence? It is this association that is the essence of the code. Today, it plays a role in translation, but we think it first arose to serve quite a different function. If so, this is an example of a common feature of evolution: structures that today serve a complex function arose first to serve a simpler one.

6.1 Modifications of the code

For many years the common genetic code (Table 6.1) was thought to be universal. Recently, some interesting exceptions have been found (Table 6.2). These are of two types: either a stop codon is used to code for an amino acid, or a codon has been reassigned to a different amino acid. At first sight it is hard to see how this could happen. To alter the meaning of a codon in one particular gene might be a selective advantage, just as any mutation might be, but to alter its meaning wherever it occurs throughout the genome must surely be disastrous. Yet it is clear from Table 6.2 that such changes have occurred. This is important for the following reason. We will see later that the code has some apparently adaptive features: if the code, once it had arisen, was unalterable, it is hard to see how this could be so.

The reassignment of stop codons to amino acids is relatively easy to conceive, since the former by definition occur only at the end of mRNA. Hence, stop codons with new sense could cause only the attachment of at most one or two new amino acid residues to the end of a protein chain. Protein synthesis could, in principle, be terminated by simple run-off: the lack of a subsequent codon could be equivalent in effect to a 'stop' command. The actual evolutionary pathway in the bacterium *Mycoplasma capricolum* seems to have avoided even this minor problem, as explained in Fig. 6.1. Similar processes have apparently occurred in ciliated protists (such as *Paramecium*) and in the mitochondria of various organisms (Table 6.2).

Table 6.1 The genetic code for nuclear genes

Codon	Amino acid	Codon	Amino acid	Codon	Amino acid	Codon	Amino acid
UUU	Phe	UCU	Ser	UAU	Tyr	UGU	Cys
UUC	Phe	UCC	Ser	UAC	Tyr	UGC	Cys
UUA	Leu	UCA	Ser	UAA	Ter	UGA	Ter
UUG	Leu	UCG	Ser	UAG	Ter	UGG	Trp
CUU	Leu	CCU	Pro	CAU	His	CGU	Arg
CUC	Leu	CCC	Pro	CAC	His	CGC	Arg
CUA	Leu	CCA	Pro	CAA	Gln	CGA	Arg
CUG	Leu	CCG	Pro	CAG	Gln	CGG	Arg
AUU	Ile	ACU	Thr	AAU	Asn	AGU	Ser
AUC	Ile	ACC	Thr	AAC	Asn	AGC	Ser
AUA	Ile	ACA	Thr	AAA	Lys	AGA	Arg
AUG	Met	ACG	Thr	AAG	Lys	ACG	Arg
GUU	Val	GCU	Ala	GAU	Asp	GGU	Gly
GUC	Val	GCC	Ala	GAC	Asp	GGC	Gly
GUA	Val	GCA	Ala	GAA	Glu	GGA	Gly
GUG	Val	GCG	Ala	GAG	Glu	GGG	Gly

Table 6.2 Changes in the universal code

Codon in universal code	Changes to	System Nuclear	Mitochondrial
UGA, Stop	Trp	*Mycoplasma*	All except plants
AUA, Ile	Met		Yeast
			Metazoans, except echinoderms
	Met to Ile		Echinoderms
AGR, Arg	Ser		Metazoans, except vertebrates
	Stop		Vertebrates
AAA, Lys	Asn		Flatworms, echinoderms
UAR, Stop	Gln	*Acetabularia* Ciliated protozoans except *Euplotes*	
CUN, Leu	Thr		Yeast
CUG, Leu	Ser	*Candida cylindracea*	
UGA, Stop	SeCys	Vertebrates, eubacteria (in special enzymes)	

From Jukes & Osawa, 1991.

Reassignment of a codon to a different amino acid has occurred in some mitochondria and the nucleus of a yeast species. For example, AAA, the code of lysine (Lys) in the common code, codes for asparagine (Asn) in the mitochondria of flatworms and echinoderms. A plausible scenario for such a change has been suggested by Ohama *et al.* (1990): as in the reassignment of a stop codon illustrated in Fig. 6.1, the change involves an intermediate stage in which, because of AU → GC pressure, the AAA codon (and UUU anticodon) are not used. Similar mechanisms were probably responsible for other codon changes as well (Osawa & Jukes, 1988, 1989).

The moral of the story is that minor reassignments in the genetic code can occur provided that some codons disappear transiently through directional mutation pressure, and then reappear with an altered assignment, while protein sequences remain unaltered. Thus, although there exist some deviations from the common code, these are only minor variations on a fairly robust theme. The near-universality of the code argues for a unique origin. The occasional variations show that codon assignments can change, albeit rarely. In the early evolution of the code, such changes may have been commoner.

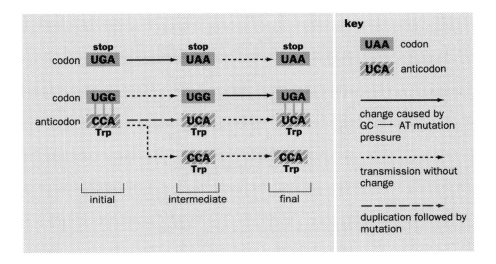

Fig. 6.1 Change of assignment of UGA from a stop codon to a tryptophan codon. Mutation pressure from GC to AT converted all UGA stop codons to UAA, without any change in proteins. At this stage UGA codons were virtually absent. The usual codon and anticodon for tryptophan are UGG and CCA, respectively (they are exactly complementary: no wobble is required). Duplication of the tryptophan tRNA was essential for the subsequent steps. The anticodon of one of them mutated to UCA, which could then read UGA codons as tryptophan. Thus, subsequent mutations of some UGG codons to UGA were allowed, with no change in protein sequence.

6.2 The origin of the code I: the top–down approach

Now we come to the origin problem proper. As with many questions in biology, one can take either a bottom–up or a top–down route of enquiry. The bottom–up strategy tries to get from elements with known properties to systems, whereas the top–down approach attempts to figure out the details by considering the pattern of behaviour at the system level. In evolution, these approaches often have an additional twist since they may refer to possible temporal order as well: rudimentary elements may precede systems not only in the structural but also in the temporal sense. Since we have a reasonably rich description of the present state of the genetic code without having ideas of commensurate clarity and detail about the earliest stages, we shall first take the top–down route. This, unfortunately, will not take us right to the bottom.

Do similar codons code for chemically similar amino acids?

Looking at Table 6.1 makes one think that chemically similar amino acids have similar codons. Such an arrangement would have two advantages. First, genetic mutations, which usually change only one base in a codon, will have less-damaging effects. Second, as Swanson (1984) argued, it would minimize the effects of translational errors. We will refer to these two effects, jointly, as load minimization.

Sonneborn (1965) argued that the present favourable pattern arose by codon rearrangements very early in evolution. His suggestion has been challenged on two grounds: first, that the suggested pattern does not exist; and second, that there is no mechanism whereby an optimal code could evolve. We will discuss these two objections in turn.

Wong (1980; see also Wong, 1988, for a review of the topic) argued that the present code does not minimize the chemical differences between amino acids specified by neighbouring codons, and that it could be improved by a series of minor rearrangements. Instead, he prefers the idea (Wong, 1975) that amino acids that are neighbours in biochemical pathways have adjacent codons. We will argue later that there is probably something in this idea, but it seems that Wong may be mistaken in thinking that the code could readily be improved from the standpoint of load minimization. Recently, Di Giulio (1989) and Haig & Hurst (1991) have reexamined the case with different results. The essence of the latter's approach is as follows. Take the common code, and, for each codon, record the change in amino acid polarity caused by substituting, one by one, the first, the second and the third codon base. Then calculate the mean squared change. Next, generate random codes with the same redundancies as the original one, and calculate again the corresponding mean squared polarity changes. It turned out that among 10 000 randomly generated codes, only two were

more conservative than the common code as regards polarity distances between neighbouring amino acids. Thus it seems that load minimization was a crucial factor in shaping the code.

An interesting observation concerns the coding capacity of complementary codons. For example, the codon UUU of Phe is complementary to codon AAA of Lys. If all such complementarities are listed, and amino acids thus related are connected, the emerging graph consists of three internally connected, externally disjoint subgraphs. As Zull & Smith (1990) point out, this results in the surprising fact that the secondary structure of a polypeptide translated from a sense nucleic acid will be conserved in the secondary structure of the polypeptide translated from the 'nonsense' (complementary) nucleic acid. Moreover, as Konecny *et al.* (1993) have found, a mutation leading to synonymous (silent) amino acid substitution in the sense strand results in a conservative substitution in the polypeptide coded for by the nonsense strand (i.e. the original amino acid will be replaced by one physicochemically similar to it). These authors suggest, therefore, that the code has evolved in the direction of optimal double-strand coding.

How should we evaluate this suggestion? To be sure, contemporary double-strand coding is extremely rare, but this does not imply that it was so in the early days of evolution. In fact, cases have been found where proteins unrelated in their sense strands turn out to be related in a sense–antisense fashion. This seems to favour the hypothesis. However, as Konecny *et al.* (1993) point out, these strange observations arise solely because the middle base of the codons primarily determines the physicochemical class of the coded amino acid (see below). If, for whatever reason, this was an outcome of some other mechanical or selective process, the observations regarding double-strand coding would follow as a byproduct or epiphenomenon.

How could the code evolve to minimize the load?

A more serious difficulty arises in imagining an evolutionary mechanism that could alter the code, once it existed, so as to minimize the load. To improve the code, some codon–amino acid assignments must be changed. But once this change has been made, many amino acids, in many proteins, will be substituted at once to a different one. Thus, if rearrangements of the code are accompanied by amino acid substitutions in proteins, it is impossible to get out of the trap of a suboptimal code.

Changes in the code, however, do not necessarily imply amino acid substitutions. Somewhat to his surprise, Szathmáry (1991c) found that such 'code-shuffling' works quite generally. Taking the current wobble rules as given, one can design scenarios whereby all the codons of certain pairs of neighbouring amino acids can be exchanged; one does not get stuck in an impasse. In effect, what is required is a succession of four events, in each of which a single codon is

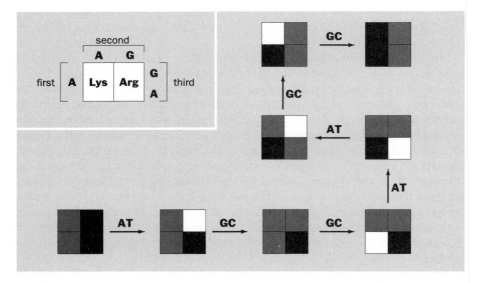

Fig. 6.2 A scenario for complete codon swapping. Initially (top left), codons AAA and AAG code for lysine, and AGA and AGG for arginine (as in the existing code). The arrows indicate substitution pressure. Starting at the bottom left, the AGG codon disappears under GC→AT pressure, and is then reassigned to lysine. A succession of such changes ends (top right) in a reversal of the initial assignments.

reassigned. Figure 6.2 illustrates the process for the particular case of lysine (Lys) and arginine (Arg). As in other examples of codon capture, alternating AT and GC pressures have been assumed. A similar scenario can be developed for the threonine–arginine pair.

Are all neighbouring codon sets exchangeable in this way? It turns out that the answer is yes, provided that the wobble rules allowed exclusive recognition of C. Nothing seems to rule out such a possibility in principle. Thus the contemporary wobble rules themselves must have arisen through evolution. Wobble depends partly on base modification, which is unlikely to be primitive: the early codon–anticodon pairing probably obeyed Watson–Crick rules. We conclude that certainly half, and perhaps all, neighbouring codon sets could have been swapped. Codon swapping, together with the simpler codon capture, could have combined to form the mechanism of general codon shuffling, a means of fine-tuning in the evolution of the genetic code (Szathmáry, 1991c).

Woese (1965) suggested an alternative process, ambiguity reduction, whereby the code could have reached its present state. It seems reasonable that the earliest analogues of contemporary aminoacyl-tRNA synthetases—let us call them assignment catalysts—could not unambiguously discriminate between chemically similar amino acids. Therefore, groups of amino acids were encoded

Fig. 6.3 Correlations between the features of the code. All correlations are significant at the 0.05 level.

by groups of codons. (In fact, no such discrimination can be perfect, just as one cannot reduce the mutation rate to zero. The current rate of mis-aminoacylation of a tRNA by the synthetase is one in 10^4 cases.) This amounts to saying that the primeval code must have been not only degenerate but also ambiguous, but meaningfully so. Later, as specificity developed, a group of similar codons shared by related amino acids was partitioned among the amino acids in a way which at least preserved, or possibly increased, load minimization. According to this view, the evolution of the code was, in Orgel's (1989) phrase, like the coming into focus of an obscure picture.

What can tRNA phylogeny tell us?

It is an old idea that comparison of tRNA sequences should shed light on the development of the genetic code (Wong, 1975). For example, there is a high degree of similarity between tRNATyr and tRNAPhe from the diatom *Scenedesmus* (Wong, 1988). This is very interesting since Tyr and Phe are biosynthetically related, although their polarities are rather different. A more systematic analysis was carried out by Szathmáry & Zintzaras (1992), using data from Eigen *et al.* (1989). The latter were able to determine the ancestral sequences of tRNAs for all the amino acids but one. They showed that 10 nucleotide positions are likely to have been very well conserved since the fixation of the code: it is these subsequences that have been used in analyses involving tRNA relatedness.

Distances can be defined between pairs of codons (by the number of point mutations), between ancestral tRNA sequences, between the chemical properties of the amino acids, and by the number of biochemical reactions between two amino acids in the metabolic map. The question is how well these four distances correlate with each other. The results are summarized in Fig. 6.3. Note that, in this analysis, ancestral tRNA sequences do not include the anticodon.

Two main conclusions emerge:

- tRNA sequence similarity correlates with polarity and biosynthetic distance. Since these are not correlated with each other, it seems that both have been important in the evolution of the code.

- Anticodon differences correlate more strongly with polarity than with biosynthetic position. This fact, together with the absence of a correlation between codon and tRNA, suggests that there has been some evolution since the original codon assignments were made. The mechanism could have been codon swapping, which destroyed any correlation between anticodon and tRNA that might have existed. Another possible explanation is that the anticodon loop is more ancient than the rest of the tRNA (Rodin *et al.*, 1993): our bottom–up hypothesis will be shown to be consistent with this idea.

'The code within the codon'

One observation about the code is compatible with the conclusions of the last section. It is striking that the first (5′) position of the codon seems to be associated with biosynthetic kinship: codons of the amino acids belonging to the shikimate, pyruvate, aspartate and glutamate families tend to have U, G, A and C in this position, respectively. In contrast, polarity seems to be associated with the middle position. This regularity was referred to as the 'code within the codons' by Taylor & Coates (1989), who suggested that it originated before the tRNA/ribosome system. Consistent with this is the older conclusion (Swanson, 1984) that the first base of a codon controls amino acid size (purine and pyrimidine meaning small and large, respectively), and the middle base controls cloisteredness (purine means external, pyrimidine means internal).

Why is the code redundant?

Redundancy of the genetic code is apparent: two amino acids (Ser and Arg) are coded for by six codons each, and family boxes (four codons for an amino acid) are quite common. Why? One suggestion is that amino acids commonly used in proteins have more codons. Is this a feature of the code that evolved because it minimizes load, or a mere consequence of (neutral) mutation pressure, once the pattern of redundancy was fixed?

It is entirely plausible that the relation between codon number and amino acid abundance should have arisen through mutation pressure. However, the correlation is not perfect (Jukes *et al.*, 1975). Since the intracellular pH is approximately neutral, one would expect the frequencies of acidic (Asp and Glu) and basic (Arg and Lys) amino acids to be equal. This is true, but not what one would expect from codon multiplicities: on that basis, one would expect 13.1 per cent basic to 6.6 per cent acidic amino acids. In order to neutralize the overall protein charge, natural selection has acted 'against the code'. It is also the case

that proteins have more Ala and less His, Cys, Pro, Ser and Leu than would be expected from codon multiplicities alone.

We conclude that assignment multiplicities were primarily determined by factors that had nothing to do with amino acid functionality in proteins later in evolution (we shall return to this later). Neutral mutation pressure has ensured that amino acids with more codons are commoner in proteins, but the correlation is not perfect: there has also been 'selection against the code'.

6.3 The origin of the code II: the bottom–up approach

How did amino acid–nucleotide associations originate?

We now come to the central question: how did specific associations between amino acids and nucleotides originate? Gánti (1983) argued that the origin of translation and the genetic code must have followed a route leading from ribozymes to protein enzymes with identical functions. We accept this argument, but it is clear that no crude picture of the process will work. The distance between two nucleotides in nucleic acids is about 0.34 nm, that between two amino acid residues in proteins is about 0.36 nm. This correspondence implies that if we have a functional ribozyme, it could not act as a messenger for an isofunctional protein using the present three-letter genetic code, since critical distances within the molecules would not be preserved.

We think that the first amino acids used regularly by RNA organisms were acting as coenzymes of ribozymes (Szathmáry, 1990, 1993a). We suggest that there existed cofactors, each consisting of an amino acid bonded to one or more nucleotides: the amino acids would increase the chemical range and specificity of the ribozymes, and the nucleotides would constitute a 'handle' whereby the cofactors were located, by base pairing, relative to the ribozyme (Fig. 6.4). It is relevant that many coenzymes today have such nucleotide handles. Thus we think that contemporary amino acid activation (that is, the binding of specific amino acids to specific tRNAs, which is the essence of the code) is a relic of the final step of an ancient process synthesizing amino acid cofactors, and that ribozymes catalysing it were the first assignment catalysts.

Amino acids have a wider variety of chemical functions than unmodified nucleotides, so it must have paid to use as many cofactor types as possible. There could be many kinds of cofactor, each consisting of a particular amino acid bound to a particular oligonucleotide. The advantage of such an assignment would be twofold: it enables ribozymes to bind these cofactors primarily by their handles, and the unambiguous assignment of amino acids to handles ensures that the right cofactor is used at the right place. Amino acid cofactors could have

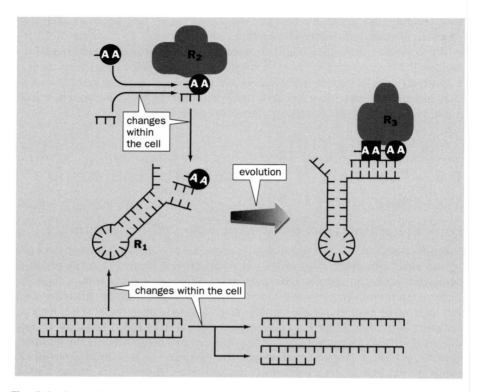

Fig. 6.4 A hypothesis for the origin of the genetic code, as explained in the text. R_1, Ribozyme with cofactor, ancestral to mRNA; R_2, ribozyme linking amino acid to oligonucleotide, the ancestral assignment enzyme; R_3, ribozyme linking amino acids, ancestral to the ribosome.

been used at critical points only; for example, to aid acid–base catalysis or the formation of hydrophobic pockets.

The reason we assume that handles were made of oligonucleotides is chemical. Binding should have been specific but reversible (transient). Theoretical considerations (Eigen, 1971) as well as recent experiments (Kazakov & Altman, 1992) indicate that a trinucleotide could have been an ideal cofactor handle. The triplet code would thus have a chemical explanation.

Of course, a given cofactor could have been used by many different ribozymes. The emergence of a specific assignment could have been aided by affinities between amino acids and nucleotides. It is known, for example, that hydrophobic amino acids in various solutions prefer the neighbourhood of A, the most hydrophobic base. Any such affinity could have been amplified by the active site of the cofactor-synthesizing ribozymes.

Since nearly every function of ribozymes could have been fulfilled better by proteins, at least in principle, longer tracts of adjacent amino acids would have been formed on ribozymes. At this stage, the analogy to modern translating machinery is clear. The original nucleotide handles must have been transformed into ancient adaptors (primordial tRNA analogues).

The next stage would have been the formation of oligopeptides. This would have had the benefit of making adaptors reusable. At this stage a differentiation of ribozyme genes is to be expected: some versions of the original gene produced RNAs that became more efficient messengers while losing catalytic ability, whereas other descendants were still involved as subunits in functional ribonucleoprotein enzymes. Such a division of labour must have been advantageous, since assembly of functioning enzymes would have been faster and messengers would have become reusable. Ribozymes catalysing peptide synthesis would have been advantageous, and become the ancestors of ribosomal RNAs. Contemporary formation of the peptide bond seems to occur in an RNA-catalysed manner (Noller *et al.*, 1992).

Thus we have got coding, messengers, adaptors and RNAsomes (primordial ribosomes). But problems remain. Perhaps the most serious is the size problem: a messenger is considerably longer than a ribozyme, and protein enzymes much longer than the short peptides that could be formed by using a ribozyme as a 'message'. The solution is not clear. It may in part depend on a feature of coding systems illustrated in Fig. 6.5. Given a triplet code (or any code with a fixed codon size), the tandem repeat of an oligonucleotide codes for a polypeptide consisting of a corresponding tandem repeat: this is true regardless of the reading

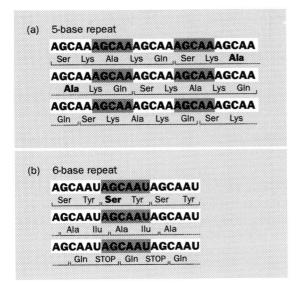

Fig. 6.5. The growth and coding capacities of oligomeric repeats (after Ohno, 1987). (a) The repeated pentanucleotide AGCAA codes for a polypeptide chain with the pentapeptide repeat Ser–Lys–Ala–Lys–Gln in all three reading frames. (b) The hexanucleotide AGCAAU can code only for dipeptide repeats, and may be aborted by terminator codons (stop). Thus, ancient RNAs are preferably made of oligoribonucleotide modules whose sizes are not integer multiples of three.

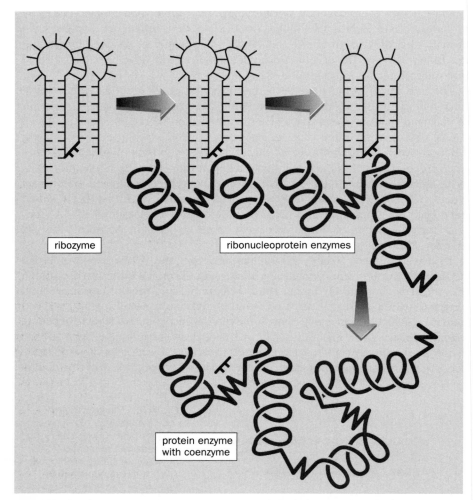

ribozyme

ribonucleoprotein enzymes

protein enzyme
with coenzyme

Fig. 6.6. Evolution of coenzymes from polynucleotides (after White, 1982). The emphasized region of the RNA molecule represents part of the active site in the ancestral ribozyme, which evolves into a cofactor of a protein enzyme.

frame in which translation takes place. For example, the repeated oligonucleotide AGCAA codes, in all frames, for the repeated peptide Ser–Lys–Ala–Lys–Gln. This helps to explain how long peptides could be coded for at a time when many triplets were still unassigned, and would therefore have operated as stop codons. Thus, merely adding random nucleotides would soon lead to termination, but short duplications would not.

The crucial assumption in our incomplete scenario is that of a continuous development of enzymatic activity: because each gene was present in several copies, experimentation with new variants was possible without the risk of losing function entirely. The original ribozyme served as a shrinking core or scaffold, onto which the novel protein structures assembled (Fig. 6.6). The ribonucleoprotein world has been a transient one, but memories of it are kept in the few contemporary examples of functional ribonucleoprotein complexes, including as important ones as ribosomes.

This imagined process implies that parts of a ribozyme can be functionally replaced by a polypeptide. There is now experimental evidence that this is possible. Some extant ribozymes come in two forms: alone or complemented by a protein. The former are longer. In one case it was found that the extra RNA and the complementing protein bind to overlapping regions of the residual RNA structure (Mohr *et al.*, 1994).

Are the codon assignments a 'frozen accident'?

The hypothesis that the code originated in the synthesis of a set of specific cofactors leaves open whether the assignments were arbitrary, or chemically determined. Crick (1968) suggested that the amino acid–codon assignments might represent a frozen accident. This does not imply that the assignments are entirely random. For example, they might be such as to minimize load, as discussed above. The hypothesis asserts only that there is no chemical reason why particular codons specify particular amino acids. For example, load minimization requires that Glu (GAA, GAG) and Asp (GAT, GAC) should be specified by neighbouring codons, but, according to Crick's hypothesis, it is a historical accident that the codons should be GAN. The fact that aminoacyl-tRNA synthetases often do not recognize the identity of the correct tRNA by the anticodon supports the idea.

This view has been challenged repeatedly. Notably, Woese (1965) coupled the ambiguity reduction hypothesis with the so-called stereochemical one in thinking that the anticodons of ancient tRNAs stereochemically recognized the cognate amino acid, as in the lock-and-key relation between substrate and enzyme. Obviously, this fit could not have led to very accurate discrimination, hence the ambiguity. According to this view, at least overall chemical similarity, if not steric complementarity, between amino acids and anticodons should be found.

Direct stereochemical match would be desirable. The most promising approach is due to Shimizu (1982), who demonstrated that there is in fact a lock-and-key relationship between space-filling molecular models of a complex consisting of the anticodon and the so-called discriminator base of tRNA (next to the 3' CCA tail), and the cognate amino acid. Measurements made *in vitro* for some of the suggested complexes show that the binding is real, although not very strong, and seems to be discriminating.

Szathmáry (1993a) proposed a detailed hypothesis for the origin and structure of the first synthetase ribozymes, which acted to strengthen the binding between these complexes and the cognate amino acid, thus increasing the degree of discrimination. An implication of his hypothesis is that the self-splicing introns found today in the tRNAs of some prokaryotes are the remnants of aminoacyl-tRNA synthetases, as first proposed by Ho (1988).

A crucial experiment to test the stereochemical hypothesis was suggested by Szathmáry (1989b). Remember the suggestion made previously that different ribozymes could be produced by RNA breeding *in vitro*. The production of RNAs able to bind amino acids seems equally possible. After sufficient experimentation, a whole array of different RNAs, each specific for a particular amino acid, can be generated. Different RNAs binding the same amino acid could also emerge. Then one could test for the significant occurrence of anticodon-like structures in the selected molecules. Until then, one is left with the feeling that the complementarity demonstrated by Shimizu (and by others for related models) could hardly have arisen by chance alone.

It is pleasant to see that Famulok (1994) has already started such a procedure. He selected RNA molecules that specifically bind to citrulline (a non-proteinogenic amino acid) and arginine. In the latter case, the binding site is codonic rather than anticodonic. One can hardly wait for a systematic test involving more and more coded amino acids.

How can translation work if it is highly ambiguous?

It is obvious that ambiguity in translation would reduce efficiency. There is, however, a non-obvious reason why ambiguity can be of special importance in the context of translation: it can lead to an 'error catastrophe' (Orgel, 1963). The idea is as follows. Suppose that, in any cell, translation errors lead to the production of malfunctioning proteins. This represents a loss of efficiency, but not a fatal one. Suppose, however, that some malfunctional proteins are themselves used in translation; for example, they are assignment catalysts. Then a single error in one round of protein synthesis could cause several errors in the next round. If so, there would be an exponential increase in the frequency of errors: an error catastrophe.

This suggestion was first made in the context of ageing, but our present concern is different: given the high ambiguity of early enzymes, could early translation work at all? The RNA world may provide a pleasant solution. As the first assignment catalysts, as well as protoribosomes, are likely to have been ribozymes rather than proteins, Orgel's malevolent error loop would not have existed, since the ribozymes (including the synthetases) would have replicated themselves without a decisive aid by newly made short oligopeptides. By the time proteins became decisive, coding would have been sufficiently accurate.

Why are there 20 amino acids?

Why we have no more than 20 kinds of amino acids is not known, but a possible answer was briefly suggested by Szathmáry (1991b), which is analogous to the hypothetical solution to the size problem of genetic alphabets, discussed in section 5.5. Increasing the size of the catalytic alphabet (the number of the different amino acids) is advantageous since the overall catalytic efficiency of metabolism increases. However, the protein synthetic machinery must discriminate between closely related amino acids in translation. It may be true that the fidelity of translation must decrease beyond a certain size of the catalytic alphabet. As translational fidelity is a less-important component of fitness than replication fidelity (proteins are disposable), one may expect that the optimum catalytic alphabet size rests at a relatively low translational fidelity and a high number of different amino acids. A detailed analysis of this problem is needed.

7 The Origin of Protocells

Cellularization has the following main aspects that we have to explain:

- The need for active (self-generated) compartmentation when metabolism is liberated from the surface.
- The origin of membranogenic molecules and membranes.
- The origin and mechanism of spontaneous protocell fission.
- The transportation problem. Simple membranes are not 'leaky' enough to permit important nutrients to pass through.
- Were the first protocells autotrophs or heterotrophs? The evolution of the first autocatalytic metabolic cycle.
- The iron-sulphur world and the RNA world: are they mutually exclusive or complementary?
- The problem of the origin of the two membranes of negibacteria, the most ancient existing group of organisms.
- The origin of chromosomes and DNA synthesis.

We shall discuss these problems in turn.

7.1 The need for active compartmentation

As we discussed before, the prebiotic pizza has the ability to localize metabolites and genes. This is advantageous for two reasons: (1) Reactants are kept in each other's proximity, which ensures that reaction rates will be high enough and that important compounds do not drift away. (2) Genes will interact, directly (e.g. by influencing each other's replication) or indirectly (by catalysing steps of metabolism), only with their neighbours: selection will thus be able to ensure cooperation among genes that would otherwise compete against each other.

Life liberated itself from surfaces a long time ago. Somehow, passive localization must have been replaced by an active process of membrane generation, maintenance and fission.

7.2 The origin of membranogenic molecules and membranes

The basic structure of contemporary biomembranes is as follows. There is a molecular bilayer of lipids, to which proteins are attached in various ways. The bilayer is formed because the membrane constituents are so-called amphipathic molecules: they have a hydrophilic head and a hydrophobic tail (Fig. 7.1). Since the binding interaction of water with itself is much stronger than that between water and hydrophobic compounds, the latter are expelled by water as much as

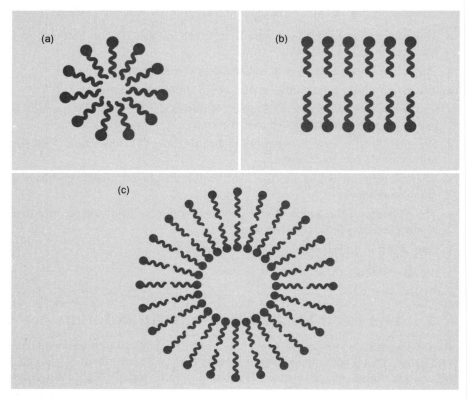

Fig. 7.1 Supramolecular structures formed by amphipathic molecules in water: (a) micelle, (b) lamelle and (c) vesicle. Note the hydrous interior of the vesicle.

possible; this results in tails coming together. A simple sheet of bilayer would still be not at the energy minimum because its edges would be exposed to water. An energetically favourable solution is the formation of a lipid vesicle. Proteins can interact either with the hydrophilic heads or with the hydrophobic tails; in the latter case, the protein must have a part consisting of hydrophobic amino acids on the protein exterior. Such a protein can be immersed into the lipid bilayer and remain there. However, for reasons given above, we think that proteins resulting from translation are latecomers to membranes.

The primordial soup hypothesis has been weak on the issue of the spontaneous formation of lipids. Once the constituents—namely, fatty acids, glycerol and phosphate are present—their spontaneous condensation is not unlikely (Eichberg *et al.*, 1977; Epps *et al.*, 1978). The problem is that membrane formation requires fatty acids with a sufficiently long linear hydrophobic chain (with more than 10–12 carbon atoms). Under plausible prebiotic broth constitutions, only

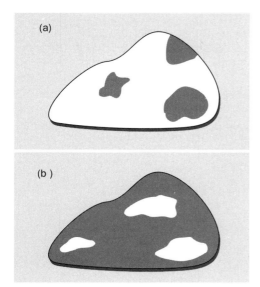

Fig. 7.2 (a) 'Oil-in-water' and (b) 'water-in-oil' patchiness on a mineral surface. The dark regions indicate surface covered by amphipathic (membranogenic) molecules. Naturally, the head groups of the molecules must be surface-binders (i.e. anionic on a pyrite surface).

branched-chain fatty acids of this size are produced, whose membrane-forming capability is uncertain. It is interesting that membranogenic, lipid-like material could be extracted from the Murchison meteorite, which suggests that some plausible abiotic pathway must be available, not necessarily under broth conditions.

Wächtershäuser (1988) called attention to the fact that the extant biochemical pathway leading to isoprenoids, from which all membranogenic molecules are formed, consists of surface-bonding chain extension steps. Once the first lipids appear in the prebiotic pizza, they change the local chemical milieu. They push water away, thus creating a more hydrophobic local environment. This makes the condensation reactions (leading, for example, to nucleic acids) more favourable, but also helps the formation of energetic anhydrids such as pyrophosphate and even ATP. This in turn helps the lipids to be equipped with strongly bonding pyrophosphate groups. Long-chain fatty acids could have formed by a variant of the archaic reductive citrate cycle on the pyrite surface.

Lipid formation results in the growth of a lipid-covered surface area (Fig. 7.2). Sooner or later, at least at some sites, 'oil-in-water' patchiness will give rise to a water-in-oil one. Further accumulation of lipids can result in semicellular structures by abstriction (Fig. 7.3). Note that metabolism of the semicell, due to the autocatalytic cycles of the pizza, is inherently autotrophic, and ions are still provided by the pyrite surface. This is the first true individuation step in the prebiotic pizza. Semicells are semi-organisms.

Later, upon the development of a sufficient transport mechanism (see below), semicells became detached from the surface to form genuine protocells. This, however, required an adequate cell division (fission) mechanism.

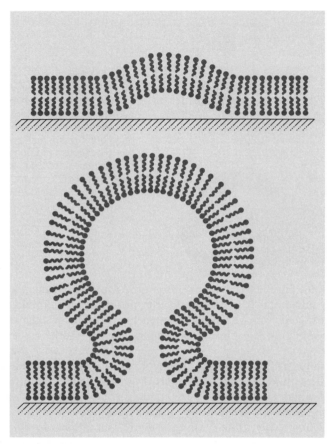

Fig. 7.3 Cellularization by abstriction from a mineral surface (after Wächtershäuser, 1988).

7.3 Spontaneous cell division

We imagine a protocell to be essentially chemoton-like (cf. Fig. 2.2): an auto-catalytic cycle must generate the membranogenic molecules inside the system. It is clear that they would spontaneously be absorbed by the existing membrane: the surface of the bilayer membrane would grow essentially in an autocatalytic manner. Two questions arise: (1) what is the consequence of synchronized autocatalytic metabolic and membrane growth and (2) is a mechanism of spontaneous division possible? We discuss these in turn.

Synchronized autocatalytic and membrane growth imply that surface area and volume increase at the same rate. As Rashevsky (1938) pointed out, this is incompatible with the maintenance of a spherical shape, since surface area and

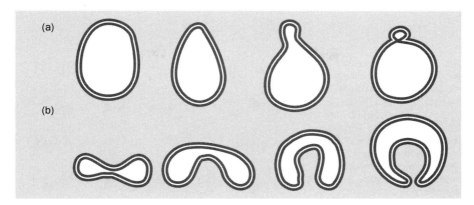

Fig. 7.4 Experimentally observed shape transformations of vesicles induced by a change in temperature: (a) 'budding' and (b) 'endocytosis' (after Lipowsky, 1991).

volume increase as the square and the cube of the radius, respectively. As the system grows, it is always possible to assign to it two imaginary spheres of unequal size whose total volume and total surface area maintain the correct ratio, but such a resolution of the conflict between surface area and volume is energetically unfavourable: the bending energy of the smaller sphere would be very high. Thus, the only feasible resolution is to divide into two when both total volume and total surface area are doubled, and two protocells of equal size can be formed (Gánti, 1978).

A more rigorous treatment is possible using results of the physics of elasticity. Let us deal first with various equilibrium membrane conformations and then with dividing systems. The basic work on the elastic properties of lipid bilayers was done by Helfrich (1973), who demonstrated the importance of curvature in controlling non-spherical shapes of membranes. The current state of these studies was reviewed Lipowsky (1991); we recapitulate some of the details.

The two crucial features are (1) sharp curvature of a membrane involves high bending energy, and (2) an edge also involves energy, arising from the partial contact between the water and the hydrophobic chains of the lipid. Provided that the total area of the membrane is large enough, the membrane can always lower its energy by forming a closed surface. If there is no constraint on the enclosed volume, the closed surface will be spherical, but if surface area and volume are both constrained a spherical shape may be ruled out.

A few experimentally detectable equilibrium shapes are shown in Fig. 7.4. The control parameter for the changes is temperature. The non-spherical shapes arise, not from any constraint on the enclosed volume, but because the inner and outer monolayers expand at different rates with temperature, perhaps because of molecular impurities.

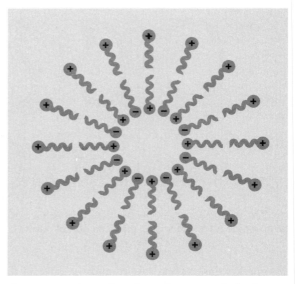

Fig. 7.5 A stable vesicle composed of membranogenic molecules with cationic (+) and anionic (−) head groups. In the inner layer, head groups are close to each other because of the ionic bond. In the outer layer, the heads are not so close because of repulsion between + charges. The two layers hence curve with opposite sign, stabilizing the vesicle.

It seems, therefore, that the formation of vesicles, their shapes, and the fission of a single vesicle into two, depend on two factors: simultaneous constraints on surface area and volume, ruling out a spherical shape and perhaps enforcing fission, and a difference in the constitution of the inner and outer monolayers. One reason for believing that mixtures of different amphiphils were important in primordial membranes is as follows. Many types of vesicle do not form without the input of considerable mechanical energy (such as sonication), and once formed they are metastable and revert to multilamellar liquid–crystalline aggregates. However, Kaler *et al.* (1989) demonstrated that stable vesicles can be formed spontaneously from a mixture of simple, single-tailed cationic and anionic amphiphils. The vesicles range from 30 to 80 nm in radius and are stable for at least 1 year. Their size depends on the total concentration and ratio of the constituents.

Safran *et al.* (1990) suggested the following explanation. Vesicles can be stabilized by the fact that the two membranogenic molecules have different concentrations in the inner and outer layer of the membrane. This allows for the two layers to have equal curvatures with opposite sign, provided the interactions between cations and anions are taken into account. A simplified, didactic scheme of this is shown in Fig. 7.5.

Fission may be explained by the fact that changes in surface area and volume drive the membrane through a series of quasi-equilibria, at the end of which the equilibrium is two equal-sized spheres. If the rate of primordial metabolism is not too high, this is a good approximation. Tarumi & Schwegler (1987) attempted to deduce an equation of protocell motion. They assumed that membrane formation is autocatalytic at the expense of a nutrient. The time evolution of curvature

can be numerically calculated: the original spherical shape is distorted to a dumbbell, and, upon enhancement of the cleavage furrow, division is supposed to follow, as a means to resolve the singularity that arises in the equations. An experimental confirmation of such a protocell division is, however, not yet available.

A promising line of experiments was initiated by Luisi and his colleagues (Bachmann *et al.*, 1992). They showed that (1) *de novo* formation of amphiphilic micelles from precursors is possible, and (2) the micelles are replicators: their density grows autocatalytically. In fact, the autocatalytic effect is so strong that it is suspected that there are two autocatalytic steps: the formation of the membranogenic molecules from their precursors is autocatalytic, and the growth of the micelle volume is also autocatalytic. The authors showed that micelles could be converted to more stable vesicles by a change in acidity by dissolved carbon dioxide. There is preliminary evidence that the vesicles can also multiply autocatalytically (Pier Luigi Luisi, personal communication). The final goal would, of course, be the synthesis of a realistic chemoton, as suggested by Gánti (1987).

7.4 The problem of membrane transport

How far does the growth and division of protocells depend on permeability of the membrane? Ideally, a system like the chemoton in Fig. 2.2 could be viable without any complications in membrane traffic. If the system is autotrophic (and it has to be, since the soup itself is 'autotrophic'), only small molecules like H_2S and CO_2 must be taken up through the membrane. Because of their small size, their diffusion through a bilayer is conceivable (Wächtershäuser, 1992).

The problem is that the idealized network of the chemoton does not deal with the ionic requirements of metabolism at all. All present-day free-living cells have active transport proteins in their membranes, pumping different ions and molecules, ultimately at the expense of ATP. For example, a Ca^{2+}/H^+ antiporter can be found in all bacteria. If one thinks that the protocell was heterotrophic rather than autotrophic, the problem is made worse by the need to import organic molecules as well through the membrane. To be sure, bilayers are rather impermeable to many electrolytes as well as to hydrophilic molecules.

The notion of a semicellular phase in evolution offers an escape. Then, the small apolar nutrients were able to pass through the membrane. In contrast, ferrous, phosphate and trace metal ions could have been provided by the mineral grain to which the semicell was attached. The surface metabolist could have spread laterally onto non-pyrite ferrous mineral grains; for example, onto ferrous sulphides with absorbed phosphate ions (Wächtershäuser, 1992). Nevertheless, semicells that could take up ions from the medium could spread to relatively ion-poor areas. At this stage any compound, small or large, that could facilitate the diffusion of ions would be inserted into the membrane and lead to a fitter

semicell. However, this would require an environment of high salinity: otherwise the ions would be lost rather than captured. Life could move into low-salinity environments only after the acquisition of active transport at the expense of ATP, presumably after the origin of translation.

The idea that facilitated transport, mediated by small molecules, is a viable possibility has been demonstrated by Stilwell (1976). He showed that bilayers become rather permeable to glycine, lysine and histidine provided simple alde-hydes are also present in the solution, the most efficient of which is pyridoxal. It should be noted that the net direction of facilitated diffusion can only follow the concentration gradient. If materials are to be carried into the vesicle, their inside concentration must be lower than their outside concentration. This condition can be satisfied if they are consumed by metabolism; for example, monomers are converted into polymers, or metal ions are adsorbed by nucleic acids or proteins. Diffusion facilitators in semicells of ferrous ions could have included lipid mole-cules with several –SH groups. Later, siderophores for proper chelation could have appeared (Wächtershäuser, 1992). Note that fragments of RNA molecules, linked to membranogenic molecules, could also have served as carriers. The main conclusion is that the origin of protocells is unlikely without the prior evolution of ion carriers in some semicellular phase.

7.5 Primordial ancestry of autotrophy

At the protocellular stage we imagine an autotrophic metabolism, aided by ribozymes. Autotrophy of the protocell is a legacy of that of the surface meta-

BOX 7A Autocatalytic cycles for CO_2 fixation

There are two extant autocatalytic cycles for CO_2 fixation: the Calvin cycle (reductive pentosephosphate cycle) and the reductive citric acid cycle. It was suggested by Hartmann (1975) that the latter may be of ancient origin. Analogous to the extant cycle one can propose an archaic version wherein the –SH groups coexist with –OH groups and the carbonyls coexist with thio-derivatives. Note the surface-bonding properties of the carboxylate (–COO⁻) and the thiocarboxylate (–COS⁻) groups. The overall balance of the proposed archaic cycle is

$$4CO_2 + 7H_2 \rightarrow (CH_2 - COOH)_2 + 4H_2O,$$

$$\Delta G = -160 \text{ kJ/mol}$$

which is exergonic by itself. Of course, H_2 is supplied by the formation of pyrite:

$$4HCO_3^- + 2H^+ + 7H_2S + 7FeS \rightarrow (CH_2 - COO^-)_2 + 7FeS_2 + 8H_2O,$$

$$\Delta G = -429 \text{ kJ/mol}$$

which is much more exergonic. It can also be shown that the proposed archaic cycle is not very 'bumpy' from a thermodynamic point of view; that is, there is no single step in it which would be highly endergonic. Thus the most ancient metabolism was

BOX 7A *Continued*

chemo-autotrophic: the energy for carbon fixation was provided by the oxidation of FeS.

An autocatalytic cycle like this is interesting in its own right, but if we want to have it as a metabolic subsystem of a (semi) cell, other products must be able to emerge from it as well. As Wächtershäuser (1990) suggested, an interesting concatenation of autocatalytic cycles is conceivable with intermediates of ever-longer carbon chains, finally leading to long-chain fatty acids, useful in lipid synthesis. Several other side-reaction classes can be explained, leading to the following products:

- the amino acids aspartate and glutamate;

- the tetrapyrrol pathway leading, for example, to porphyrins; and

- the formation of phosphate ester from geminal thiol-phosphate groups.

There are two important questions to consider about this archaic cycle: its origin and its disappearance by further transformations. We discuss them in turn.

It does not seem far-fetched to propose a slow pyrite-pulled reductive carboxylation of carbon dioxide to oxalate (via thiocarbonate), which in turn may further be reduced and carboxylated to oxaloacetate (Wächtershäuser, 1990).

The archaic cycle consists of surface-bonded intermediates. During the evolution of semicells to cells, liberation of these molecules from the pyrite surface must have happened, possibly along the following lines:

1. Catalysts with lipophilic and cationic groups appeared.

2. Special reactions to eliminate the surface-bonding groups appeared.

3. Metabolism relies on two different nicotinic amide coenzymes: NAD^+ and $NADP^+$. Their different roles in metabolism can be explained as follows. Transition from surface to cytosol metabolism aided by ribozymes may have proceeded only stepwise. It would have been extremely harmful to lift off all the surface-bonded coenzymes at once, since most reactions would have still proceeded on the surface. Coenzymes having various forms with different surface-bonding propensities are more likely to have survived this transition: while $NADP^+$ was still surface-bonded, NAD^+ was becoming used in cytosol mechanism. Ever since then (with a few exceptions) NAD^+ is used in catabolism and $NADP^+$ is used in anabolism. The origin of fermentation was possibly bound to the cytosol, hence it is a relative latecomer in biochemical evolution (see also Cavalier-Smith, 1987c).

The further evolution of the archaic reductive citric acid cycle by surface detachment is conceivable (Wächtershäuser, 1990), not forgetting that the appearance of polyanionic catalysts with thiol groups ($-CH_2-SH$), such as coenzyme A, is decisive; and that ultimately enzymes with iron-sulphur clusters, such as ferredoxins, must have appeared in metabolism as a complete (and final) replacement to pyrite.

bolist. Earlier, the idea that metabolism was originally heterotrophic was rather fashionable. This idea overlooked two important aspects: (1) the self-suffocation of unevolved heterotrophic compartments because of the lack of sufficient membrane permeability; (2) the possibility of unevolved autocatalytic cycles. If autocatalytic cycles can be formed spontaneously, they will grow in the medium. A protocell internalizing such a medium is better off than another that is not doing so: successful protocells must have been autotrophic, assimilating inorganic carbon.

The nature of the first metabolic cycle, deduced from the surface theory, has been suggested by Wächtershäuser (1990). His ideas are outlined in Box 7A.

7.6 Metabolism in ribo-organisms: the iron-sulphur world meets the RNA world

Clearly, there is a big gap between ribozymes and iron-sulphur protein enzymes. Sulphur is conspicuously absent from RNA. Therefore, one may think that proteins were absolutely necessary for the above-detailed step of changeover from surface to cytosol metabolism. That is not so. Benner *et al.* (1987, 1989) list the functional groups available for RNA and protein enzymes, including secondary chemical modifications as well as cofactors. They show that an RNA-based redox metabolism is not impossible. They suggested certain crucial biochemical processes that may have originated in a ribo-organism. In all of them RNA plays some role even today. In identifying such processes, we should look for the following lines of evidence:

- The trait must be found at least in eubacteria, but possibly in archaebacteria and eukaryotes as well.
- There should be some aspects of the process that serve no selected function in extant biochemistry, and which therefore can best be explained as relics of an ancestral condition.

Application of the last criterion needs knowledge-based intuition in organic chemistry. For example, pyrophosphate in some cases is as good a phosphate donor as ATP or, occasionally, S, S-dimethylthioacetate is as good a methyl donor as the RNA coenzyme S-adenosylmethionine: hence, the involvement of adenine in extant biochemistry may be an ancestral relict.

Bearing these criteria in mind, it is possible to suggest a number of catalysed metabolic processes that may well have been present in ribo-organisms, before the evolution of translation and informed protein enzymes: these are listed in Box 7B.

BOX 7B Metabolic processes in ribo-organisms

DNA replication. This is somewhat surprising, but almost obvious. Homology of DNA-dependent RNA polymerases for all organisms is likely. Ribonucleotide reductases are present in all organisms. Their substrates are ribonucleotides, even though alternative chemical routes to deoxyribonucleotides would be possible. Some ribonucleotide reductases use the RNA cofactor B-12. Last but not least, DNA synthesis is much easier than translation, and would by itself confer a high selective advantage through an increase in replication fidelity and in resistance against hydrolysis. (A complication here is that there are at least three different, non-homologous ribonucleotide reductases in the living world. It can thus be suggested that the ancient reductase was a ribozyme, which was replaced by proteins relatively late in evolution.)

Ribo-organisms synthesized tetrapyrrols. The C5 pathway for the synthesis of the tetrapyrrol precursor 5-aminolevulinate involves reduction of an RNA-bound glutamic acid. Moreover, this pathway is used to synthesize chlorophyll in photosynthetic eubacteria and eukaryotes, and to synthesize factor F430 and B-12 in archae-bacteria.

Ribo-organisms were presumably photosynthetic. Besides the above observations, this is supported by noting that photosynthesis, like DNA synthesis, is much simpler than translation, and confers a high selective advantage over chemosynthesis inherited from the prebiotic pizza. We suggest that some iron-sulphur RNAs were operative in this phase of evolution (a missing link between the iron-sulphur world and present-day iron-sulphur proteins in electron transport).

Ribo-organisms synthesized terpenes. Higher terpenes are synthesized along similar routes in all organisms. Most remarkably, the eubacterium *Rhodopseudomonas acidophila* contains hopanoid-RNA conjugates in its membrane, where RNA serves as the polar head of the molecule (this finding is consonant with the view that RNAs could have played a role in early facilitated diffusion and in the control of membrane composition; see above). The acceptance of terpene biosynthesis removes the need for fatty acid biosynthesis in ribo-organisms. The latter requires biotin (non-RNA coenzyme) and an acyl-carrier protein. This does not prove that fatty acid synthesis is a latecomer; in fact, we have seen above that it can be deduced from surface metabolism.

7.7 The evolution of specific enzymes

Contemporary enzymes are highly specific for a given reaction: true bi-functionality is an extreme exception. This raises two questions: (1) what was the ancestral state of the enzyme system and (2) how did evolution proceed from

that state to the current one? These questions were considered by Kacser & Beeby (1984).

Imagine, for the sake of simplicity, a series of simple biochemical reactions, the steps of which can be catalysed either by mono- or multi-functional enzymes. There is empirical evidence that unprogrammed polypeptides have weak and non-specific catalytic activity. For this and other reasons, the initial state was probably a set of multifunctional enzymes. These would have been of low efficiency, because the active sites of early enzymes could not be expected to provide a good and specific fit for all reactions in which they participated. This reasoning applies equally to ribozymes and to protein enzymes. Under what conditions is evolution to a set of efficient, specific enzymes possible?

One might expect mutational improvements to be available, but this is not so if every gene in the cell has a copy number of one, essentially because increased specificity for one reaction will result in diminished efficiency in others. The analysis by Kacser & Beeby (1984) suggests that, with a copy number of one for most genes, the system is sitting in a mutational trap from which escape is difficult.

The situation is changed when gene dosage is allowed to vary. As the gene dosage for a particular gene, and hence the enzyme concentration, is increased, there is a diminishing return in terms of an increase in metabolic flux. If allowance is made for the cost of a higher gene dosage, there is an optimal dosage for each gene. For a system that has evolved to this optimal gene dosage, there is an escape from the mutational trap described in the last paragraph. The system is now free to experiment, by mutating one copy of a gene. Kacser & Beeby (1984) found that, given mutation, the end state of such a system was a set of highly specific enzymes. The crucial elements of the process are the duplication and divergence of genes, leading to an efficient division of metabolic labour between enzymes.

7.8 The origin of the two negibacterial membranes

A picture of a strange protocell, the inside-out cell or obcell, was formed by Blobel (1980) and extended by Cavalier-Smith (1987c). The main merit of the obcell idea is that it can explain the origin of the two membranes (a double membrane as opposed to a single molecular bilayer) of Gram-negative bacteria simply by flattening, bending and lip fusion of the original obcell (Fig. 7.6). Metabolic communication of the two membranes would have been possible by the newly formed Bayer's patches—contemporary contact regions between the two membranes, where the cell wall is missing. The evolution of porins, preformed protein 'holes' in the outer membrane to let through all molecules with a weight less than about 600, would have happened during this transformation

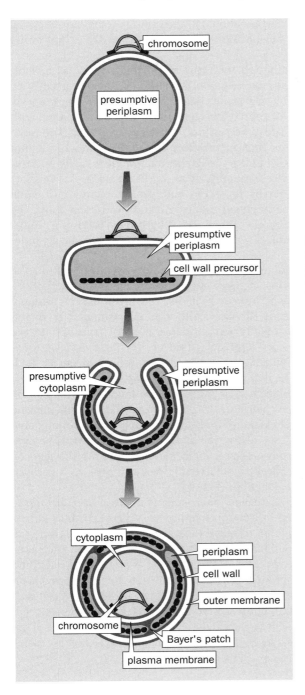

Fig. 7.6 The origin of the negibacterial cell through obcell formation, bending and lip fusion (from Blobel, 1980; Cavalier-Smith, 1987c).

(Cavalier-Smith, 1987c). The explanation of the origin of the negibacterial membranes is important since negibacteria form the most ancient extant group of life.

It might be the case that, although the Gram-negatives seem to be ancient, their second membrane evolved relatively late, after the origin of protocells, by some unknown mechanism. If, however, we take Cavalier-Smith's point of view and argue that the formation of the double membrane was concomitant with that of the protocell itself, an adequate scenario must be developed. The main problem with the obcell idea is that metabolites and gene products are not localized well enough. We suggest that if the double membrane is really ancient, it must have originated in the era dominated by semicells on surfaces. We show two topological possibilities whereby protocells with double membranes could have been formed: either by semicellularization on, and abstriction from, the pyrite surface of a double membrane (Fig. 7.7); or by the collision of a single-membraned protocell with a semicell. One should choose the biochemically more plausible option.

We prefer the long-term evolution of a double-membraned semicell on the surface, for the following reasons: it takes advantage of the pyrite surface for localization and nutrient supply just as well as Wächtershäuser's original semicell with a single membrane; it allows the gradual formation of a presumptive cell wall to increase rigidity; but, because efficient transport to the inside of the semicell becomes advantageous (see above), insertion of (functionally) porin-like molecules into the outer membrane will be favoured; and, finally, contact sites between the two membranes (presumptive Bayer's patches) are likely to appear. Those semicells possessing a more favourable membrane structure would spread to new surfaces through metabolic and membrane growth more rapidly than others lacking it. Since the supply of membranogenic molecules to the outer membrane is also required, Bayer's patches would be used for this directional traffic of membranogenic molecules coming from the semicell interior. Abstriction of such semicells would readily result in protocells with an envelope resembling that of negibacteria (Fig. 7.7).

It should be noted that multilamellar, often spherical, structures of amphipathic molecules were observed in electron microscopes long ago; their formation does not need the participation of membrane proteins or cell-wall material (Bangham & Horne, 1964). Similarly, Kaler *et al.* (1989) observed the spontaneous formation of two vesicles, one in the inside of the other. The division mechanism of a double-membraned protocell could still have followed that governed by bending energy and curvature, provided the presumptive cell wall was not rigidly sealed yet, but consisted of patches of wall pieces instead. However, this would have called for a very tight control of the composition of the four membrane layers (two bilayers). This idea needs to be checked by calculations.

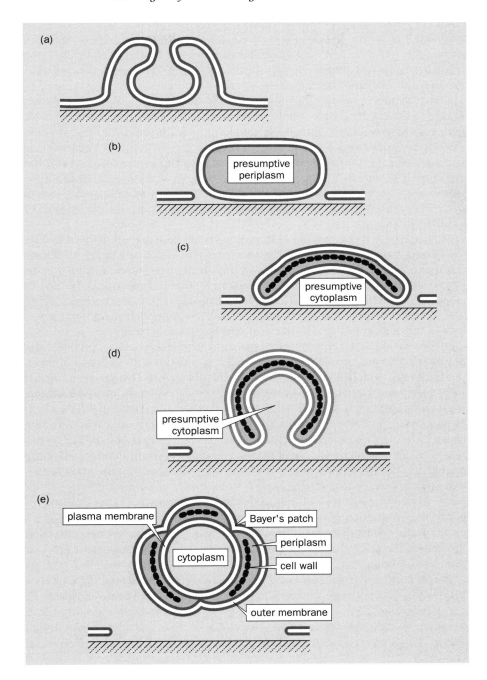

Fig. 7.7 Origin of the negibacterial cell on a mineral surface; the genetic material is not shown.

7.9 The origin of chromosomes

During the evolution of protocells, unlinked genes must have given rise to linked ones. It is thus essential to discern the selective force as well as the molecular mechanisms enabling this transition to take place.

A mathematical model (Maynard Smith & Szathmáry, 1993) suggested the nature of the selective force. The starting point is the stochastic corrector model with its unlinked genes. Imagine a simple protocell with two essential genes. Under what conditions does a primitive chromosome, with the two genes linked, increase in frequency in the population? First, one has to take into account that chromosomes will have a competitive disadvantage within the cell, since it takes longer to replicate them than unlinked genes. The exact magnitude depends on several conditions, but a twofold replicative disadvantage for the chromosome is a good conservative estimate.

This assumption of a twofold disadvantage needs justification: if there were no such disadvantage, the origin of chromosomes would not be a problem. Thus, if the number of replicase molecules working on a chromosome (i.e. two linked genes) were twice that working on solitary genes, there would be no disadvantage. However, we think this idea is mistaken (Fig. 7.8): chromosomes will indeed take longer to replicate than isolated genes. The question then is: how is this within-cell disadvantage of chromosomes overcome?

It turned out that chromosomes out-competed isolated genes, provided that (1) both genes are needed for efficient cell reproduction and (2) the number of molecules per protocell is small. The reason is that when two genes are linked, complementation in the daughter cell is ensured. One way of understanding what is happening is to take a 'gene's-eye' view. Let A and B be two complementary genes, and AB the chromosome. Is an A gene better off on its own, or as part of a chromosome? On its own, it replicates faster within a cell. However, an A gene that is part of a chromosome is certain to find itself, in the next generation, in a cell that also contains a B gene: that is, it is certain to be in a fit cell, whereas an isolated A gene may find itself in an unfit cell, with no B gene. A second advantage of linkage is the synchronization of replication, and hence the elimination of within-cell competition between different genes. This is particularly important if one of the genes replicates much faster than the other: the greater the difference in within-cell fitness of different genes, the stronger the selection for linkage.

The molecular scenario for the origin of linkage is complicated (Szathmáry & Maynard Smith, 1993a). Solutions for three problems have been suggested: (1) the origin of an RNA chromosome linking RNA genes; (2) the origin of transcription (of linked genes); and (3) the origin of DNA replacing RNA chromosomes. Because of the many biochemical details involved, we briefly discuss just (2) and (3); (1) is conceivable through linking of genes either by a ligase ribozyme or through copy-choice recombination, when the copying of one gene is

made continuous with that of the other through an exchange of the templates by the replicase. The crucial problem is, of course, how such an event could allow undisturbed functioning of the whole genetic system.

The regulation of replication and the origin of transcription rests on an assumption by Weiner & Maizels (1987) in their 'genomic tag' model: replicable molecules must have had a tRNA-like tag at their 3′ end, which was recognized by the replicase (working in the 5′ → 3′ direction). RNA viruses replicated

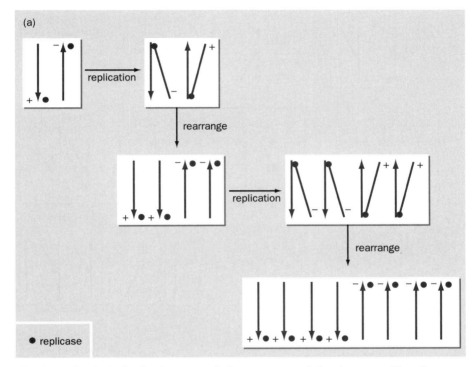

Fig. 7.8 The intrinsic disadvantage of chromosomes relative to genes. The diagrams (above and next page) show the replication of (a) genes and (b) chromosomes. The assumptions in both figures are: (1) The rate-limiting step is the time taken for the replicase to travel along a gene. Each arrow marked 'replication' takes the same time. The arrows marked 'rearrange' involve the separation of copies from templates, and the binding of a new replicase: these steps take negligible time. (2) A new replicase can bind when the previous one has travelled the length of a gene. The conclusion would be the same if, for example, a new replicase could bind when the previous one had travelled half the length of a gene, but the diagram would be more complicated. The conclusion is that chromosome replication is a slower process. Even if we take the state marked * as the starting point, after two replication steps the initial configuration is not even tripled, whereas in (a) it is quadrupled. The essential reason is that a newly synthesized strand cannot act as a template until its synthesis is complete.

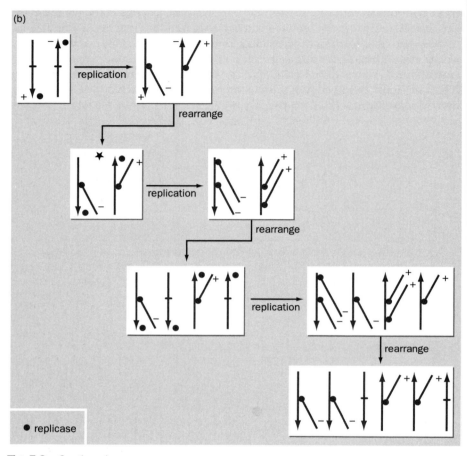

Fig. 7.8 *Continued*

without a DNA intermediate all have such tags. Accepting this assumption, the dynamics of replication and gene expression is as follows (Fig. 7.9). Bear in mind that the RNAs in question are all single-stranded, hence there are + and − copies. By definition, let + be the gene and − be the ribozyme (the product of 'gene expression'). Now, imagine that the tag of the + strand is a stronger replicase target site than that of the − strand. This would result in an over-production of − strands. Another enzyme, RNAse P, is assumed to liberate the molecules from their tags. We assume that RNAse P liberates the − strand more effectively than the + strand. RNAse P is a known enzyme from bacteria: it cleaves off the 5′ unused region of pre-tRNAs; its active part is a ribozyme. Of course, RNAse P cannot be too effective, or all strands would become un-

replicable. The result of these mechanisms would be a protocell with a higher concentration of enzymes than of genes, as required for efficient metabolism.

The origin of DNA chromosomes must have been facilitated by a special type of reverse transcriptase, copying both RNA and DNA into DNA. Interestingly, the reverse transcriptase of the AIDS virus is exactly like this. It might be thought that the main selective advantages of having a DNA genome is its increased stability (e.g. through the repair of single-stranded lesions) and decreased mutability (through the possibility of mismatch repair). However, all this would be possible in the case of double-stranded RNAs. Therefore we believe that the increased chemical stability of deoxyribose and T (as opposed to ribose and U, respectively) was the primary selective force for the takeover by DNA. The main reasons why we think that RNA preceded DNA are: protein synthesis can proceed without DNA but not without RNA; RNAs can act as ribozymes; many coenzymes still have nucleotide handles; deoxyribonucleotides are always synthesized by reduction of ribonucleotides (Lazcano *et al.*, 1988).

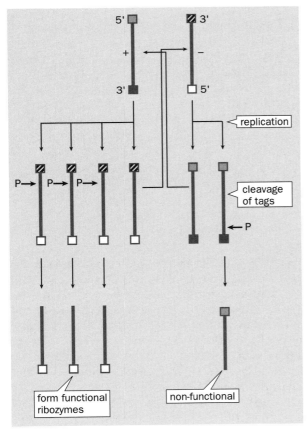

Fig. 7.9 Replication and 'transcription' of RNA genes. + and − strands are copied from each other. They have 'genomic tags' (dark-stippled and hatched squares) at their 3′ ends, differing somewhat in sequence, so that the + strand is a better target for the replicase. The strands must carry sequences complementary to the tags of the opposite sign strand at their 5′ ends (open and light-stippled squares). The enzyme RNAse P liberates the tags from the rest of the sequence. The former are eliminated by a specific nuclease, and the latter are free to function as ribozymes. Usually, only one of the two complementary strands will be functional. If the − stand is the ribozyme. its overproduction (because + is a better template) is advantageous.

To sum up, we suggest that:

- Semicells and some protocells were chemoautotrophic.
- Some protocells became photoautotrophic.
- Protocells were essentially ribo-organisms.
- DNA became the genetic material in ribo-organisms, before the evolution of translation and protein enzymes.
- Linkage of genes was strongly selected for in protocells.
- Protocells divided autonomously as a consequence of internal metabolism.
- Protocells might have inherited a doubled bilayer membrane from the ancestral semicells.

8 The Origin of Eukaryotes

8.1 The problem

The basic structures of a bacterial and a eukaryotic cell are shown in Fig. 8.1. The differences whose origins call for an explanation are as follows:

• The bacterial cell has a rigid outer cell wall, usually made of the peptidoglycan, murein. In eukaryotes, the rigid cell wall is not universal, and cell shape is maintained primarily by an internal cytoskeleton of filaments and microtubules.

• Eukaryotic cells have a complex system of internal membranes, including the nuclear envelope, endoplasmic reticulum and lysosomes.

• Bacteria have a single circular chromosome, attached to the rigid outer cell wall. In eukaryotes, linear chromosomes are contained within a nuclear envelope, which separates transcription from translation: communication between nucleus and cytoplasm is via pores in the nuclear envelope.

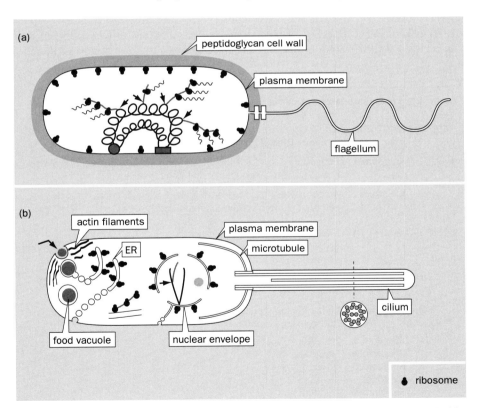

Fig. 8.1 The structure of (a) a prokaryotic and (b) a eukaryotic cell (after Cavalier-Smith, 1988). ER, endoplasmic reticulum; →, replicase.

- Eukaryotes have a complex cytoskeleton. The actomyosin system powers cell division, phagocytosis, amoeboid motion and the overall contractility to resist osmotic swelling. Microtubules and the associated motor proteins (kinesin, dynein and dynamin) ensure the accurate segregation of chromosomes in mitosis, ciliary motility and the movement of transport vesicles. Intermediate filaments form the structural basis for the association of the endomembranes and nuclear-pore complexes with the chromatin to form the nuclear envelope, while other intermediate filaments help to anchor the nucleus in the cytoplasm.

One crucial difference between prokaryotes and most eukaryotes has been omitted from Fig. 8.1: this is the presence of mitochondria, and, in plants and algae, of chloroplasts. The reason for the omission is that, on the scenario for eukaryote origins that seems to us most plausible, these intracellular organelles originated later in time than the structures shown in the figure.

The differences between these cell types justifies the recognition of two empires of life (above the kingdom level): Bacteria and Eukaryota (Cavalier-Smith, 1991a; Table 8.1). (It is interesting that this taxonomic rank was recognized by Linnaeus.) Within each of the empires, there are two major categories: Bacteria

Table 8.1 A classification of living organisms

EMPIRE BACTERIA	Cyclic DNA bound to cell envelope; ribosomes 70S; no nucleus, endomembranes, cytoskeleton or cytosis
Kingdom 1	Eubacteria (acyl ester lipids, as eukaryotes); includes negibacteria, spirochaetes, togobacteria and posibacteria
Kingdom 2	Archaebacteria (isoprenoidal ether lipids; no murein); includes sulphobacteria, methanobacteria and halobacteria
EMPIRE EUKARYOTA	Nucleus with linear DNA attached to nuclear envelope; cytoskeleton, endomembranes, cytosis; cilia with 9 + 2 structure usual
Superkingdom 1	Archaezoa—ribosomes 70S; no mitochondria, peroxisomes and chloroplasts
Superkingdom 2	Metakaryota—ribosomes 80S; mitochondria and peroxisomes; chloroplasts common
Kingdom 1	Protozoa
Kingdom 2	Chromista
Kingdom 3	Fungi
Kingdom 4	Plantae
Kingdom 5	Animalia

Shortened from Cavalier-Smith, 1991a.

consist of the kingdoms Eubacteria and Archaebacteria, and Eukaryota are divided into the superkingdoms Archaezoa and Metakaryota.

The justification for these divisions is as follows. The Archaebacteria, in contrast to the Eubacteria, never have murein cell walls, and their single cell membrane contains isoprenoidal ether rather than acyl ester lipids. Analysis of sequences suggests that these two groups branched off from each other rather early (Woese & Fox, 1977). The name of Archaebacteria was meant to indicate their alleged ancestry relative to Eubacteria. This interpretation has been challenged for some time, and recent sequence analyses indicate that the Archaebacteria are indeed specialized and secondary, rather than ancient and primary creatures. The correct evolutionary relationships among the Eubacteria, Archaebacteria and Eukaryota cannot be determined by analysis based on comparison of a single molecule from the variety of extant species, because the root of the tree cannot be determined. A sensible way to overcome this difficulty is to compare duplicated genes, the duplication of which happened before the divergence of the three lineages (Miyata *et al.*, 1991). If both genes are present in all three lineages, then it is possible to make composite phylogenetic trees that include both sequences. Such composite trees were made for two different protein pairs: the proton-pumping ATPase non-catalytic and catalytic subunits, and the ribosomal elongation factors EF-Tu/G. The composite tree for the former pair is shown in Fig. 8.2. It consists of two subtrees of similar topology, which

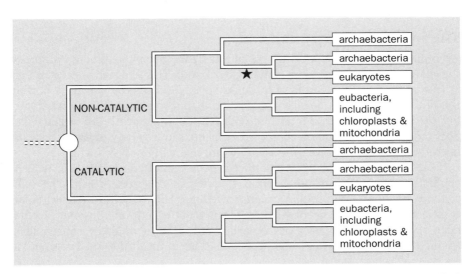

Fig. 8.2 The phylogeny of the catalytic and non-catalytic subunits of ATPase (simplified, from Miyata *et al.*, 1991). If only data on the non-catalytic subunit were available, it would be possible to construct a tree with * as a root, with the Archaebacteria as the ancestral group (although such a tree would imply very unequal rates of evolution). But the whole tree, with both subunits represented, does not permit such an interpretation.

are joined together by a line segment within which the original duplication must have occurred. The position of this segment within the subtrees locates their root. In all cases investigated the root is on the eubacterial branch. Also, it is apparent that archaebacteria are more closely related to eukaryotes than to eubacteria: the former two are evolutionary sister groups. Based on such evidence, even Woese now thinks that archaebacteria are less ancient than eubacteria (cf. Pool, 1990).

Within the Eukaryota, the cells of the Archaezoa and Metakaryota are rather different. Mitochondria, chloroplasts, peroxysomes and hydrogenosomes are absent in the former. Also, archaezoan cells have a *trans*-Golgi network, but no flattened Golgi dictyosomes. Most important, their ribosomes are 70S in size, as in prokaryotes. It is very unlikely that their lack of mitochondria is secondary, since free-living species also lack them. They cannot metabolize oxygen at all to generate ATP; presumably, all of them rely on glycolysis only. Recent sequence data (Vossbrinck *et al.*, 1987; Sogin *et al.*, 1989) are consistent with the view that the Archaezoa are the basal, paraphyletic group from which other eukaryotes evolved. Table 8.1 summarizes the latest state of macrotaxonomy.

In the next section, we suggest a possible scenario for the origin of the eukaryotic cell: in essence, this follows the suggestions of Cavalier-Smith (1987a and earlier). The different stages are then discussed in more detail.

8.2 A possible scenario

The loss of the outer cell wall

According to Cavalier-Smith, the crucial event that precipitated the evolution of the eukaryotic cell was the loss of the rigid cell wall. This loss made necessary, or at least made possible, a series of secondary changes. But why was the cell wall lost in the first place? A possible explanation is that some competing prokaryote invented an antibiotic that interfered with peptidoglycan synthesis, as penicillin does today. Existing bacteria cope with antibiotics in a variety of ways, but our ancestors may have evolved the ability to survive without a cell wall.

If so, the first requirement would be the evolution of an endoskeleton which could resist osmotic and other forces that would otherwise disrupt the cell. Actomyosin is able to prohibit osmotic swelling of the cell, and microtubules resist compressive and shearing forces. The loss of the cell wall, however, would open up new adaptive possibilities as well. In particular, it would become possible for the cell to engulf particles—phagocytosis. This in turn led to the evolution of a system of internal membranes.

The organization and transmission of the genetic material

One crucial requirement for all cells is a mechanism whereby, at division, one complete copy of the genetic material is transmitted to each daughter cell. In

bacteria, the segregation mechanism depends on the attachment of the chromosomes to the cell wall. With the loss of the cell wall, a new mechanism, mitosis, was needed. Although mitosis was evolved in the first place because the bacterial mechanism would no longer function efficiently, once it existed it made possible a second change that was necessary if a further increase in complexity was to follow. In bacterial replication, there is typically only a single origin of replication for the whole chromosome, although in damaged cells DNA replication can initiate at other sites (Magee *et al.*, 1992). The significance of this latter observation is hard to evaluate: we are not aware of evidence that normally dividing bacteria can have multiple origins. Current ideas about how the bacterial chromosome segregates at cell division (see Fig. 8.4) would seem to rule out multiple origins, but these ideas are still controversial. The matter is important for the evolution of complexity, because, if it turns out that prokaryotes really are limited to a single replication origin, this in turn limits the total amount of DNA that can be replicated. In eukaryotes, there are multiple origins of replication, allowing an indefinite increase in total genomic DNA.

At the same time, the evolution of a nuclear envelope separated transcription from translation, with important consequences for genomic organization.

The symbiotic origin of organelles

Once phagocytosis had evolved, there were ample opportunities for endosymbiosis. In prokaryotes, symbiosis, whether mutualistic or parasitic, involves the introduction into the cell of DNA molecules—plasmids or phages. Eukaryotes can have endosymbiotic cells. That mitochondria and chloroplasts are descended from originally free-living prokaryotes is no longer in doubt (Margulis, 1970, 1981; Gray & Doolittle, 1982; Gray, 1989). Much more controversial is the claim (Margulis *et al.*, 1979; Margulis, 1981, 1991) that eukaryotic microtubular structures—cilia and the mitotic spindle—are derived from a spirochaete-like endosymbiont. This represents the major alternative to the scenario outlined above. In both theories, the origin of the cytoskeleton is crucial, but Cavalier-Smith sees it as an endogenous response to the loss of the cell wall, whereas Margulis thinks that a symbiotic event was the crucial step. This question, and the general role of symbiosis in evolution, are discussed in section 8.6.

The symbiotic origin of certain cell organelles was first suggested by Russian biologists (Mereschkowsky, 1910; Kozo-Polyanski, 1924); the first of them coined the term symbiogenesis itself.

8.3 The origin of intracellular membranes

In phagocytosis, a solid food particle is enclosed in a food vacuole within the cell. This has obvious advantages over the alternative mechanism of producing extracellular enzymes and absorbing the products of digestion through the cell

surface: neither the enzymes nor the digestive products can diffuse away. Phagocytosis requires that a spherical vacuole be budded off from the inside of the plasma membrane that bounds the cell. This process, and its inverse—the fusion of a vacuole with the plasma membrane, or the fusion of two vacuoles—have led to the development of a complex system of internal membranes.

In bacteria, membrane proteins are synthesized by ribosomes attached to the plasma membrane, and inserted directly into the membrane. In eukaryotes, new membrane is synthesized by ribosomes attached to the rough endoplasmic reticulum (ER). Vacuoles are budded off the ER, and then incorporated into the plasma membrane, bringing about growth. Note that the rough ER carries attachment sites for ribosomes, and includes enzymes synthesizing the cell-wall lipids, whereas these components are missing from the vacuoles that are budded off to contribute to the plasma membrane. This may have been the first differentiation of types of cell membrane to evolve, but many further elaborations have since appeared.

Lysosomes, capable of producing digestive enzymes internally, are also budded off the ER. If a lysosome fuses with a food vacuole budded off from the plasma membrane, the contained food will be digested. In evolution, however, if the first eukaryotes were descended from a prokaryote performing extracellular digestion, the first food vacuoles budded off would be able to secrete enzymes internally.

The ER was probably derived in evolution from vacuoles invaginated from the plasma membrane, and flattened to form ER cisternae. The fusion of a number of such cisternae then formed the nuclear envelope. This mode of origin would explain why the nuclear envelope consists of an inner and outer membrane, joined at the nuclear pores, and why the outer membrane carries ribosomes, and is continuous with the ER.

8.4 The origin of mitosis

Bacteria have a single circular chromosome. Replication starts at a single origin of replication, and proceeds in both directions round the circle, ultimately meeting at a unique terminus. In contrast, eukaryotes have a number of linear chromosomes, each with many origins of replication. Eukaryotes, therefore, can have a much larger total quantity of DNA. In the most rapidly dividing bacterial cells, division takes about 20 minutes, whereas it takes about 40 minutes for the replicon to travel from origin to terminus. Hence, two replicons must be travelling along the chromosome at any one time. It follows (Fig. 8.3) that the cell contains four copies of genes near the origin, but only one copy of genes close to the terminus. In a slowly growing cell, however, there will be only one copy of most genes most of the time. This gene-dosage problem places a limit on the number of replicons that can simultaneously be active. If there is also a limit on the rate at which the replication complex can travel along the chromosome, this places a limit on the size of the total genome.

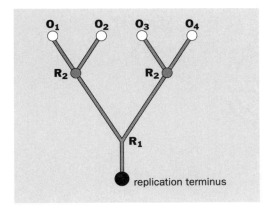

Fig. 8.3 Replication in a prokaryote. During rapid growth, a second round of chromosome replication starts before the first is completed, so that four copies of a gene close to the origin are present, but only a single copy of a gene close to the terminus. R_1, R_2, first and second replication forks.

No comparable limit seems to exist for eukaryotes. For example, in the early development of *Drosophila*, when nuclei are dividing but there is no growth and no cell division, nuclear division takes about 20 minutes. Since the further evolution of prokaryotes may well have been prevented primarily by the limitation on genome size, the limitation to a single origin of replication is of some importance. The most likely explanation is that the bacterial mechanism of chromosome segregation would not work if there was more than one origin. With the origin of mitosis this limitation disappeared. Of course, the change did not take place because it would, in the future, permit an increase in genome size: the immediate cause was the loss of the rigid cell wall. Our account of the change follows closely that given by Cavalier-Smith (1987a).

The mechanism of chromosome segregation in bacteria is illustrated in Fig. 8.4. Although the details are still to some degree speculative, the essential features are:

- The chromosome is attached to the cell wall at the origin (O) and at the terminus (T).

- As the two replication forks travel along the chromosome, they carry the new origin (O) with them.

- When replication is complete, the new origin is attached to the cell wall, at the opposite end of the cell from the old origin.

- The terminus is released from the cell wall, and re-attached half way between the two origins: the movement may be brought about by chromosome coiling.

- The terminus replicates. Growth of new cell wall between the new and old terminus then completes cell division.

In contrast, the familiar mitosis, found in higher eukaryotes, depends on the formation of a bipolar spindle from microtubules, emerging from the two centrosomes, one situated at either pole of the spindle. Chromosomes are attached to

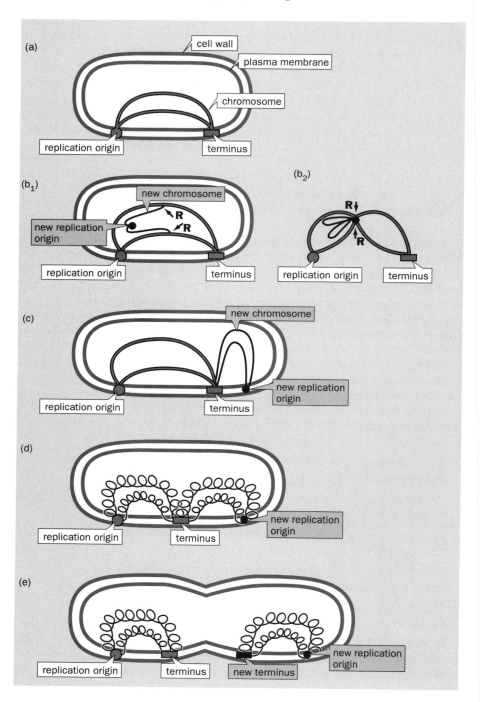

the equator of the spindle by their centromeres, and, after replication, the daughter chromosomes are pulled by their centromeres to the two poles. This process is so different from that illustrated in Fig. 8.4 that it is hard to see how one could have evolved from the other. A clue is provided by a form of mitosis, 'pleuromitosis' (Raikov, 1982), found in some primitive protists. This will be described in more detail below (Fig. 8.6).

A possible intermediate between bacterial division and pleuromitosis is illustrated in Fig. 8.5. It resembles bacterial division in the following features: (1) the chromosome is circular and (2) replication starts at a single origin, and proceeds in both directions. It differs in that the new origin, instead of being carried by the replicon to lie next to the terminus, is separated from the old origin by the action of the cytoskeleton and lies next to it, attached to the plasma membrane. Cell division is then completed by a furrow between the new and old origin.

In the bacterial system, chromosome segregation depends on the new origin being carried by the replicon to lie next to the terminus, attached to the cell wall. This in turn requires that a replicase travels the full distance from origin to terminus. It therefore rules out the existence of subsidiary origins, since, if they did exist, the replicase starting at the primary origin would travel only as far as the next subsidiary origin, and not to the terminus. In the mechanism illustrated in Fig. 8.5, however, there is no barrier to the appearance of additional origins of replication.

The transition from this hypothetical mechanism to pleuromitosis is illustrated in Fig. 8.6. Note the following points:

- The origin of replication—or, strictly, the structure attaching the origin, initially to the cell wall and later to the plasma membrane—is homologous to the centrosome: the terminus is homologous to the centromere.

- The circular chromosome is converted into a linear one, with a centrally placed centromere, first by breaking the circle at the origin and then by detaching the free ends from the centrosome. The free ends become telomeres. Although initially there would be only one acrocentric chromosome, there would be no barrier to the acquisition of additional chromosomes. One possible advantage of the change illustrated in Fig. 8.6 is as follows: in a circular chromosome, sister chromatid recombination prevents proper disjunction (Fig. 8.7).

Fig. 8.4 (*opposite*) The segregation of bacterial DNA. (a) The cell before replication. (b_1) Replication proceeding in both directions. The origin and replication complexes are shown exploded: in fact, the new origin is bound to the replicases, as shown in (b_2), and so is carried to the terminus. (c) Replication is complete, and the new origin is attached to the cell wall. (d) The terminus is reattached midway between the new and old origins. (e) The terminus divides, and the formation of a septum leads to cell division. R, replicase. (After Cavalier-Smith, 1987b, based on Jacob *et al.*, 1963, Sonnenfeld, 1985, and others.)

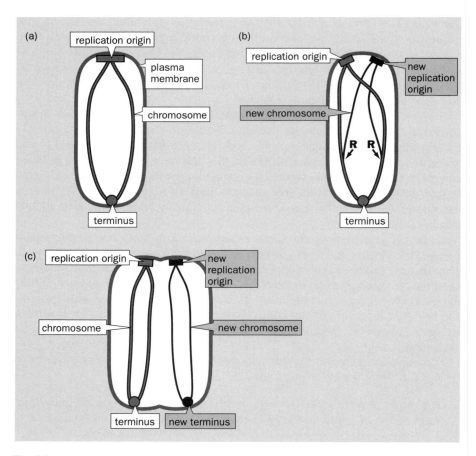

Fig. 8.5 A hypothetical mechanism of chromosome segregation intermediate between the bacterial mechanism (Fig. 8.4) and pleuromitosis (Fig. 8.6c). (a) Cell before division. (b) Replication proceeding in both directions: new and old origins have moved apart. (c) Chromosome replication is complete: a furrow is initiating cell division. R, replicase.

Pleuromitosis (see Fig. 8.6) is characteristic of some primitive protists. Typically, the two half spindles lie side by side, as shown, and the centrosome and centromere are attached to the inside of the nuclear membrane, not to the plasma membrane. Since, however, the nuclear membrane is an invaginated portion of the plasma membrane, this presents no difficulties.

Finally, the formation of a bipolar spindle (Fig. 8.6) requires the relative rotation of the two half spindles, and the cross-bonding of two microtubules of opposite polarity to form a spindle fibre.

In the light of some recent data, this picture of bacterial genome replication, and hence of the transition to the analogous eukaryotic mechanism, should be

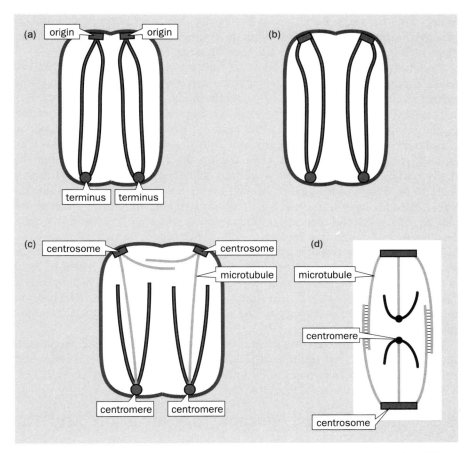

Fig. 8.6 A possible history of mitosis. (a) The hypothetical mechanism shown in Fig. 8.5. (c) Pleuromitosis. (d) Modern bipolar mitosis. In (b), the original circular chromosome has been broken at the origin, and in (c) the free ends are no longer attached to the centrosome, which is the homologue of the structure that attached the origin of replication to the plasma membrane.

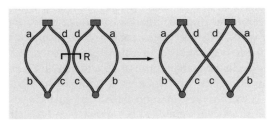

Fig. 8.7 Recombination between circular chromosomes: a,b,c,d represent successive regions of a circular chromosome. Sister-strand recombination at R prevents chromosome disjunction.

modified in two ways: bacteria can have more than one chromosome, and replicated bacterial chromosomes seem to be partitioned with the aid of mechanochemical proteins. We discuss these findings in turn.

Campbell (1993), reviewing non-standard bacterial systems, collected evidence that bacteria may have two or three chromosomes. The genome of *Rhizobium meliloti* consits of a 3.4-megabase chromosome and two large plasmids of 1.4 and 1.7 megabases. There is no evidence that these plasmids are dispensable, so they could as well be called chromosomes. *Rhodobacter spheroides* has two chromosomes, of 3 and 0.9 megabases. Thus it seems that there are no constraints preventing prokaryotes from having several chromosomes: after all, many have several plasmids, often with an active partition system. But they do seem to be limited to a single replication origin per chromosome: even the few linear plasmids have a single origin. To have two chromosomes instead of one can at most halve the replication time, whereas to have 10 replication origins would speed up replication tenfold.

There is evidence that bacterial chromosomes are not separated only by the growth of the cell membrane–cell wall complex, as originally suggested by Jacob *et al.* (1963). Mechanochemical proteins seem to be active during fission, both in chromosome positioning and in septum formation (Hiraga, 1993). The proteins involved show some homology to the eukaryotic mechanochemical proteins, myosin, dynamin and tubulin.

In conclusion, although prokaryotes may have more than one chromosome, they have not evolved multiple replication origins on a single chromosome perhaps because the movement of the terminus towards the origin, shown in Fig. 8.4, is needed for successful partition.

8.5 The nucleus, genome organization and the origin of introns

Introns: gene shuffling or fossil transposable elements?

The evolution of the nuclear envelope separated transcription from translation. This had several consequences for the subsequent evolution of the eukaryotic genetic machinery, of which the most controversial is the presence of introns in protein-coding genes. Two explanations of introns have been proposed. The first (Gilbert, 1978) is that the exons represent ancestral genes that once coded for peptides that functioned on their own. Existing genes consist of several such genes that have been brought together to code for a single protein. On this hypothesis, the introns mark the boundaries between the ancestral genes. There are many cases in which a eukaryotic gene with introns is homologous to a bacterial gene without them. On Gilbert's view of such cases, the introns are primitive, and have been lost from bacterial genes. There are a few cases in which it is certain that complex proteins have been built up from simpler

components in this way (for a review, see Blake, 1985). However, if Gilbert's idea is true in general we would expect to find:

- The region of a protein coded for by a single exon should occupy a single region in space, and should perform a single function: there is argument about how often this is true.
- A few bacterial protein-coding genes should still have introns: this seems not to be the case.
- In some cases, there should be genes in prokaryotes corresponding to single exons in eukaryotes: no such examples are known.

The observational support for Gilbert's idea is therefore weak. There are also theoretical difficulties. It is hard to imagine how the precise removal of so many introns could have happened. More serious, the ribosomes attach to bacterial mRNA immediately the transcript separates from the DNA template. It is hard to see, therefore, how introns could be spliced out before translation: the result would be the synthesis of erroneous proteins.

A recent analysis by Stoltzfus *et al.* (1994) provides rather strong evidence against the exon-shuffling theory. They asked whether the exons of four proteins, including alcohol dehydrogenase and triose-phosphate isomerase (TPI), which have been cited in favour of Gilbert's hypothesis, do in fact encode discrete units of protein structure. For example, considering the 14 introns of TPI, they found that the average distance of their insertion points from the boundaries of protein secondary structures is 5.9 base pairs, which does not differ significantly from the random expectation of 6.5 base pairs. It seems that exons do not correspond to ancient peptide pieces.

The second, and in our view more plausible, hypothesis (Cavalier-Smith, 1987a) is that introns are fossil transposable elements that invaded the genome after the evolution of the nuclear membrane, and have since lost the ability to transpose. Probably the first introns were self-splicing: they would have been efficient and relatively harmless intragenomic parasites (Orgel & Crick, 1980; Doolittle & Sapienza, 1980). Later, the loss of self-splicing ability would have evolved alongside the evolution of splicing enzymes coded for by the host genome.

The taxonomic distribution of introns

In order to assess the evolutionary history of introns more closely, it is advisable to look at their macrotaxonomical distribution. The major intron classes are as follows (Rogers, 1990):

Group I: these introns are often self-splicing and mobile. This may be possible by reverse self-splicing, but many of them encode proteins that enable them to transpose at the DNA level.

Group II: can be self-splicing, but the consensus sequences and the exact reaction mechanism are different from those of the previous group. Many of them encode a reverse transcriptase-like protein that may help them spread by insertion into novel sites. Group II self-splicing introns have been found in cyanobacteria and the purple non-sulphur bacteria (proteobacteria) — that is, in the ancestors of plastids and mitochondria, respectively (Ferat & Michel, 1993).

Group III: in these introns, the exact splicing mechanism is unknown.

In addition, there are two other categories.

Spliceosomal introns: occur only in pre-mRNA of eukaryotic (metakaryotic) nuclei; most eukaryotic introns are of this type. The spliceosome consists of proteins as well as RNA, and carries out the splicing process.

Protein-spliced introns: occur only in certain tRNA and rRNA genes.

It is rather disturbing that we do not know about the intron composition of archaezoans at all. Eubacteria seem to have only self-splicing tRNA introns, and nothing else. (The self-splicing mRNA introns in *E. coli* T-even phages are likely to have been picked up later in evolution.)

The evolution of introns

Based on the pattern of intron distribution, the following scenario can be drawn (Cavalier-Smith, 1991b):

* Group I and II self-splicing introns may originate from the RNA world. It is unlikely that they have ever widely invaded protein-coding genes of eubacteria, since translation of incompletely spliced messengers must have been very detrimental. It is to be expected that group I–II introns will be found in archaebacteria.

* Since protein-spliced introns occur in archaebacteria and in the nucleus of metakaryotes, they probably originated in the common ancestor of archaebacteria and archaezoans.

* The origin of spliceosomal introns is the most interesting. It may have occurred in the ancestors of archaezoans, after the origin of the cell nucleus. The source may have been group II self-splicing introns: there is similarity in the splice-signals as well as in a peculiar splicing intermediate—the so-called lariat. In a well-isolated compartment such as the nucleus, it is possible that most group II introns undergo deletions, and become dependent on the splicing activity (in *trans*) of complete introns. The spliceosome is likely to be a descendent of a group-II intron–protein complex. Because of the barrier posed by the nuclear membrane, incompletely spliced products do not generate chimaeric proteins.

The presence of self-splicing mRNA introns in the essentially prokaryotic translation system of plastids and mitochondria seems at first sight to invalidate the argument that chimaeric proteins translated from incompletely spliced mRNAs cannot be tolerated. The possible escape from this contradiction is that self-splicing is faster than spliceosomal splicing anyway, so that the chance of such a translation is reduced, and that because of the drastically reduced size and the intracellular multiplicity of these genomes, the consequence of a small chimaeric protein population is not so harmful as it would otherwise be. As it happens, mitochondria possess a few chimaeric proteins, some of which have even acquired some function.

A consequence of the previous scenario is that spliceosomal introns must be of late origin. This suggests that they, or their group II ancestors, must have been inserted into many protein-coding genes. Although this seems to be easier for the former (using reverse transcription), we have evidence for insertion of truly spliceosomal introns as well. The most economical explanation for the pattern of discordantly positioned introns across taxa in genes of serine proteases, collagens, actins and tubulins is insertion (often relatively late). Many authors have tried to explain some of these patterns by means of multiple removals and shifts of introns, but these explanations are very unlikely since often they must suppose a huge number of initial introns, often separated by only a few nucleotides.

There is also evidence for intron loss; processed pseudogenes are very good examples. Reverse transcription is a necessary component process in such losses. Organisms with relatively small genomes seem to have lost more introns than others; *Drosophila* and yeast lack certain introns in actin, triose-phosphate isomerase and glyceraldehyde-3-phosphate dehydrogenase (GAPDH) genes, which are present both in plants and animals.

Intron movement is a very special case: there is only a single well-established example of it, in the carbonic anhydrase gene. As such movements require double frameshifts, they are very improbable, which acounts for their extreme rarity.

A crucial implication of the foregoing is that eubacteria have never had introns in protein genes. The distribution of introns, as well as mechanisms for insertion, make this view rather credible. The pattern of introns in genes of chloroplast and cytosol GAPDH, however, remains a mystery. Some of the introns are at identical positions. Although both genes reside now in the nucleus, recombination between them cannot explain the pattern, since the plastid gene is entirely prokaryotic in the vicinity of the splice sites. It was concluded that 'intron existence predated the divergence of eukaryotes and prokaryotes' (Shih *et al.*, 1988). However, lack of any similar evidence for other genes makes us cautious.

Palmer & Logsdon (1991) reanalysed this case. They found that only two of the introns were at identical positions. In any case, there are difficulties about dates. Plastids are perhaps 1000 million years old, and the common ancestor of

plastids and plants 3000 million years old. Hence the assumption of common ancestry requires that some groups of bacteria retained the introns for 2000 million years, transmitted them to plastids, and then lost them. The authors conclude that the two introns are the result of parallel insertions.

Finally, a brief reflection on 'exon shuffling'. There is widespread evidence for this in extracellular proteins (such as immunoglobulins) in multicellular animals. This shows that the process is possible, but it does not resolve the issue of whether genes were originally assembled from exonic modules or not. For exon shuffling theory to be true, exons ought to correspond to distinct protein modules. The fact is that most such modules were 'identified' with hindsight, apparently on 'aesthetic' grounds. The only convincing examples are the introns of haemoglobin and triose phosphate isomerase genes (for a review, see Go, 1991). Modules in these cases were identified by statistical means.

In conclusion, we cannot be sure that self-splicing introns were not involved in the assembly of protein-coding genes, although the conspicuous lack of corresponding introns from these genes in eubacteria suggests that they were not. On the other hand, some such introns may be discovered in the future; lack of evidence is not the evidence of lack.

A symbiotic origin for the nucleus?

The old idea that the nucleus may be yet another transformed symbiont (e.g. Mereschowsky, 1910) has recently been revived. Sogin (1991) was puzzled by the fact that the rRNA tree, unlike protein trees, places eukaryotes at the root, with eubacteria and archaebacteria as derived sister groups, whereas the protein data suggest that eukaryotes and archaebacteria are sister groups. This, he argued, could be explained if the nucleus is a chimaera: rRNA genes come directly from the ancestral 'progenote', and protein genes from an engulfed archaebacterium, which later evolved into the nucleus. We do not find this convincing, because it is based on a single molecule, 16S-like rRNA. It is more plausible that this molecule underwent rapid evolution when the 70S ribosome evolved into the 80S variant (Cavalier-Smith, 1991a).

Lake & Rivera (1994) favour another idea, that of symbiosis between two bacteria, when a Gram-negative eubacterium engulfed an archaebacterium. This would explain the striking similarity between the HSP70 protein of the archaezoan *Giardia* and Gram-negative bacteria. But a different interpretation is possible. At least two prokaryotic genes have been transferred to eukaryotes in different lineages (Smith *et al.*, 1992): these include the Fe-superoxide dismutase and aldolases. This process must not be confused with the widespread transfer of genes from organelles to the nucleus (section 8.6). Since a major innovation of the Archaezoa was their phagotrophy, there may well have been transfer of the HSP70 gene from ingested Gram-negatives to the proto-eukaryote nucleus.

We conclude that the evidence for a symbiotic origin of the nucleus is at present weak, and, in contrast to an autogeneous origin, lacks a sensible scenario.

8.6 The origin of mitochondria, chloroplasts and microbodies

The symbiotic origin

It is now generally accepted that both mitochondria and chloroplasts are descended from free-living prokaryotes (Margulis, 1970, 1981). Both contain circular DNA genomes, and independent protein-synthesizing machinery—polymerases, tRNA and ribosomes. The ribosomes resemble those found in bacteria, and not those in the eukaryotic cytoplasm: for example, they are sensitive to tetracycline but not to cyclohexamide, and their rRNA sequences are more similar to those of prokaryotes. Chloroplast DNA can be transcribed by RNA polymerase from *E. coli*, to give mRNA that can be translated by the *E. coli* protein-synthesizing machinery.

The double membrane of plastids and mitochondria was once thought to be composed of an inner bacterial, and an outer eukaryotic (residual phagosomal) layer. The ultrastructure and function of the outer membrane suggest, however, that it is a remnant of the Gram-negative outer membrane; this is in accord with the fact that cyanobacteria and purple non-sulphur bacteria belong to this group.

Changes in the genetic system of organelles

There have been changes in the genetic system of both mitochondria and chloroplasts since they became symbionts. Many of their genes have been transferred to the nucleus: the DNA content of the genome of plant chloroplasts ranges from 120 000 to 200 000 base pairs, and of animal mitochondria from 16 000 to 19 000 base pairs. In some cases, data are available to show that nuclear genes coding for mitochondrial proteins are more similar in sequence to prokaryotic than to eukaryotic genes. There is even more dramatic evidence of transfer of genes between compartments. Whole mitochondrial genomes have been found in the nucleus, and there are some chloroplast genes in the mitochondrion (Stern & Lonsdale, 1982). Mitochondrial tRNA genes have been found in the chloroplast.

There has also been evolution of the genetic machinery of the organelles themselves. In mitochondria, there are only 22 tRNAs, many of which recognize any nucleotide in the third position. More surprising, there have been changes in the meanings of some codons. For example, UGA, which is a stop codon in the universal code, specifies tryptophan in animal and yeast mitochondria, although

it continues to be a stop codon in plant mitochondria. Similarly, CUA codes for leucine in the universal code, and in animal and plant mitochondria, but for threonine in yeast mitochondria (see the section about the genetic code).

The transfer of genes from the organelle to the nucleus requires that proteins synthesized in the cytosol be transported to the correct location in the organelle: otherwise the latter would not function. This is ensured by the attachment to the protein of a 'transit peptide', which is recognized by a receptor in the organelle membrane. Hence the attachment of a sequence coding for the transit peptide to the N-terminal end of the gene is essential for the expression of the protein in the organelle. The deletion of an organelle gene could not occur until this attachment had been made to the corresponding gene in the nucleus (Fig. 8.8).

We do not know for certain how such sequences were acquired by some 700 mitochondrial proteins encoded by the nucleus. There are common features, especially a very positive initial region, with the amino acids Arg and Lys, but there is no consensus sequence, so it is likely they evolved independently. This seems surprising at first sight, but it turns out that it is not very difficult to produce a functional transit peptide. Baker & Schatz (1987) pasted randomly chosen pieces of *E. coli* and mouse DNA in front of protein genes, and found that 2.7 per cent of the bacterial inserts and 5 per cent of the mammalian ones were successful transit peptides.

We think that the most plausible scenario for the origin of transit peptides is as follows. The coding sequences originated by the duplication of the eubacterial sequence specifying the ribosome-binding Shine–Dalgarno sequence. This is an untranslated leader of the mRNA that binds to the 16S rRNA at the initiation of translation. The sequence is UAAGGAGGU, or something similar: such a region, perhaps with a few mutations, could code for the basic and hydroxylated amino acids—Arg (GGN, AGR), Lys (AAR), Ser (UCN, AGY) and Thr (ACN). Genes with duplicated coding regions for these mRNA leaders could have appeared, perhaps by recombination. If such a gene was transferred to the nucleus, translation in the cytoplasm would have started at the first Shine–Dalgarno sequence: duplicated copies of this sequence could have resulted in presumptive transit signals.

But why should many genes have been transferred to the nucleus? There does not seem to be any conspicuous physiological sense in the particular protein-coding genes that have been retained. The transfer of genes to the nucleus has made for efficiency, both in reducing the number of copies that must be replicated (two copies per cell, as opposed to many organelles per cell, each with many copies of the genome), and in enabling the specific biochemical tasks of the organelle to be carried out without interference.

It seems that it is not essential for the replication of mitochondria that they contain any genes. Some *petite* mutants in yeast turn out to involve the loss of the whole mitochondrial genome, yet the mitochondria in such cells continue to replicate, although, of course, they do not synthesize ATP.

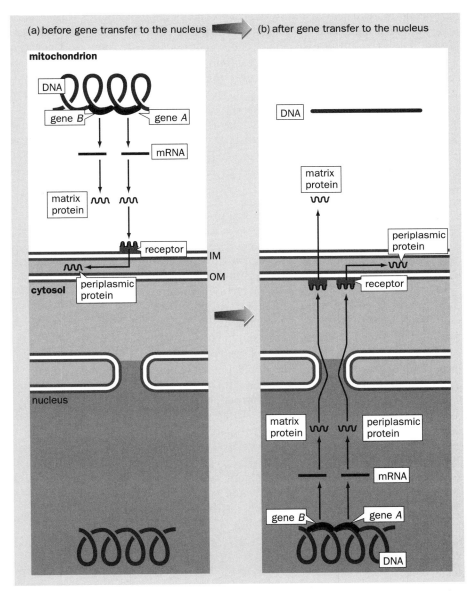

Fig. 8.8 The origin of the mitochondrial protein import system for genes transferred to the nucleus (after Cavalier-Smith, 1987d). IM, inner membrane; OM, outer membrane.

This raises the question why all organelle genes have not been transferred to the nucleus. One possible reason is that, once any change in the code had taken place, further transfer of genes to the nucleus would lead to the synthesis of

erroneous proteins. Given that any protein-coding genes remained in the orga-nelle, it was also necessary to retain the genes coding for tRNA and rRNA, as has in fact happened. If this explanation is correct, it is partly a frozen accident that particular protein-coding genes were retained: they are those genes that hap-pened to be still in the mitochondrion when the code changed. However, al-though changes in the code can account for retention of genes in the mitochondrion, it cannot account for the retention of genes in the chloroplast, since in the chloroplast the universal code is still used. Note also that we do not know much about the departure of protistan mitochondrial codes from the common code.

Peroxisomes

Peroxisomes (a common type of microbody) have a single bilayer membrane and contain various enzymes, including different peroxidases, and those for the beta-oxidation of fatty acids. Their matrix proteins are inserted post-translationally, as in mitochondria and plastids, in contrast to the lysosomes, which use the co-translational rough ER signalling and importing mechanism. They multiply by growth and division. Cavalier-Smith (1987d) suggested that peroxisomes, too, have symbiotic origins. Their ancestors were presumably Gram-positive bacteria, and have, therefore, only one membrane envelope. Oxidation of fatty acids would have added to the efficiency of phagotrophy.

The lack of DNA in these organelles needs explanation. Heijne (1986) sug-gested that the reason why functional (as opposed to *petite*) mitochondria re-tained some of their proteins is likely to be the difficulty or impossibility for certain proteins to change their conformation in a way that would allow them to be pulled through two membranes. If this plausible argument is in fact valid, then it is easy to see that presumptive peroxisomes were much less subject to such a constraint: it is easier to develop transfer of proteins through a single membrane than two.

If peroxisomes are indeed of symbiotic origin, one would expect that some proteins of Gram-positive bacteria would show homology to peroxisomal pro-teins. It will be interesting to see how far this expectation will be fulfilled.

Mutualism or slavery?

The advantage reaped from symbiosis by the host cell became possible only through the insertion into the organelle envelopes of specific 'tapping proteins', (taps) permitting the passage of metabolites from organelle to cytosol: the ade-nine nucleotide carrier for mitochondria and the organic phosphate carriers for plastids. These proteins are always encoded by the nucleus. Since their function does not need any prior gene transfer from the organelles, they are likely to have evolved from host proteins.

In any case, the evolution of taps must have taken some time. What was the nature of the association before that? It may have been parasitic, but in this case non-symbiotic host cells would have had a selective advantage over symbiotic ones. Although this does not by itself preclude the prolonged existence of the association, we suggest that there may have been a very straightforward way to turn the parasites into slaves: by intracellular prudent predation. Proto-meta-karyotic host cells may have kept protomitochondria as humans keep pigs: for controlled exploitation. Ultimately, this rudimentary type of utilization could have given way to the more elaborate metabolic exploitation by the use of taps.

We used the word 'slave' above deliberately. Although endosymbionts are often regarded as mutualistic, it may be very difficult to test this assertion. It is easy to demonstrate that the association is beneficial to the host, but impossible to prove the opposite, since free-living organelles are absent, and thus the increase in survival, growth rate and fecundity is impossible to measure (Douglas & Smith, 1989). To say that organelles benefit from the association because they die without the host cell is to confuse obligate dependence with benefit. Thus, we are inclined to think that endosymbiotic organelles are encapsulated slaves.

How many origins?

It is interesting that the ancestors of mitochondria and chloroplasts—purple non-sulphur bacteria (such as *Paracoccus denitrificans*) and cyanobacteria, respectively—are capable of both photosynthesis and respiration. This raises the possibility of capturing only one of them and using it for both purposes. There would, however, be snags. Purple bacteria cannot photosynthesize in the presence of oxygen, so they are useless as light-harvesting organelles of aerobic cells. Cyanobacteria use the same electron transport chain for both functions; separate control is thus compromised. Also, they lack a complete Krebs cycle. Thus, it is understandable why mitochondria lost photosynthesis and chloroplasts lost respiration (Cavalier-Smith, 1987d). A related question concerns the lack of eukaryotes with chloroplasts but no mitochondria. A common explanation is that the acquisition of mitochondria preceded that of plastids by some 500 million years (Margulis, 1981). But there is a more immediate explanation, based on natural selection: photosynthetic organisms with mitochondria would outcompete those without, because the latter would be inefficient respirers.

Although there is no convincing evidence for the polyphyletic origin of mitochondria, chloroplasts seem to have been adopted more than once. Some time ago, Mereschowsky (1910) suggested that all major algal groups arose independently, with the symbiotic uptake of different photosynthetic organisms (cf. Raven, 1970). More recent studies, including molecular phylogenetic analysis, indicate that the history of plastids is indeed complex (for recent reviews, see Douglas *et al.*, 1991; Martin *et al.*, 1992; Palmer, 1993). The most plausible scenario is that the origin of all plastids is ultimately monophyletic, from relatives

of the cyanobacteria. Some eukaryotes, however, have acquired their photo-synthetic organelles, not directly by engulfing a prokaryote but by engulfing an already photosynthetic eukaryote. Thus it is thought that the plastid of euglenids is an engulfed green alga, and the plastids of chromists and dinoflagellates are engulfed rhodophytes. This is particularly convincing in the case of the Chromista, whose plastids are surrounded by four membranes.

8.7 The origin of centrioles and undulipodia

Are microtubules of symbiotic origin?

The suggestion that the motility of eukaryotes with undulipodia (cilia and flagella) originated symbiotically was first made by Kozo-Polyanski (1924). This idea has been developed and vigorously defended by Margulis (1970, 1981).

In approaching the problem, one must appreciate that undulipodia and bacterial flagella are so different from each other that no homology whatsoever can be postulated. There is, however, good homology between certain eukaryotic microtubular organelles; in particular, basal bodies (kinetosomes) and centrioles are practically the same. Both are microtubular organizing centres, the first for microtubules (MTs) of undulipodia and the latter for MTs in the mitotic spindle. Thus, if one can explain the origin of either, it is easy to imagine that of the other by subsequent evolution.

The major experimental impetus to the symbiotic hypothesis is given by the finding that motility symbioses between prokaryotic and eukaryotic cells do exist. In particular, spirochaetes attached to the cell membrane can propel the host cell, since they respond to external stimuli and seem to beat in synchrony. This has led Margulis (1991) to suggest that spirochaetes evolved into undulipodia, after an initial ectosymbiosis with an archaebacterium-like host.

The evidence is still contradictory. It is encouraging that MT-like structures were searched for and found in large *Pillotina* spirochaetes (Margulis *et al.*, 1978). Antibodies to brain tubulin do bind to a tubulin-like protein in spirochaetes. It is discouraging that their polymerization is not inhibited by colchicine, and that, instead of forming MTs 24 nm in diameter, they form only fibres 5–7 nm in diameter. It is encouraging that some reports seem to indicate that centrioles and kinetosomes have essential ribonucleoproteins (RNPs) in them (reviewed by Peterson & Berns, 1980), but discouraging that, despite some attempts to show (most recently by Hall *et al.*, 1989) that they contain DNA, evidence based on immunoelectron microscopy seems to indicate rather the contrary (Johnson & Rosenbaum, 1991). This does not by itself deny the finding that undulipodial genes seem to reside on a circular linkage group (ULG) (Ramanis & Luck, 1986). Note, however, that molecular hybridization studies revealed that this linkage group is linear rather than circular (Hall *et al.*, 1989). It is also discouraging that, despite previous claims, there seems to be only one

ULG chromosome per haploid genome, and there is at least one non-undulipodial gene on it (for a review, see Johnson & Rosenbaum, 1991). There is some sequence resemblance between amino acid residues 116–134 in three tubulins and a portion of the tubulin-like protein of spirochaetes, but there is as yet no evidence for widespread homology.

The lack of DNA is not a proof for an autogenous origin, as the case of peroxisomes shows. Szathmáry (1987) has suggested a detailed scenario for the transition, according to which MTs originated from phagocytosed spirochaetes, but the modern 9 + 2 structure evolved later.

Are centrioles replicators?

Whether or not centrioles and microtubules originated symbiotically, there remains the question—are they replicators? That is, can a new centriole arise only in the presence of a pre-existing one? If so, are they 'hereditary replicators', in the sense defined in section 4.2: are there differences between centrioles that are transmitted in replication? Unfortunately, it is easier to ask these questions than to answer them.

Eukaryotic cells contain a wide variety of structures made of the protein, tubulin: these include the microtubules of the cytoskeleton, cilia and flagella, and the mitotic spindle. Eukaryotic cilia and flagella have a highly characteristic cross-section consisting of a ring of nine double microtubules, and two single axial ones: they bend actively by the ATP-driven sliding of one microtubule past another. They differ from the flagella of prokaryotes, which are single rods made of the protein flagellin, rotated by a molecular motor embedded in the plasma membrane. We will therefore use the term 'undulipodia' for the eukaryotic structures, which are not homologous to prokaryotic flagella.

Microtubules are composed of two proteins, α-and β-tubulin, coded for by nuclear genes that are homologous throughout the eukaryotes. The α and β genes are also homologous to each other, but differ from one another more in a single species than do the α genes, or the β genes, across the eukaryotes. A protein with some degree of amino acid homology to tubulin has been found in a spirochaete (Hinkle, 1991).

Most animal cells contain a centrosome, consisting of two cylinders (centrioles), embedded in electron-dense material, the microtubular organizing centre (MTOC), from which microtubules are nucleated. Centrioles are short tubes that resemble undulipodia in having a ring of nine tubulin rods, but differ in that each rod is composed of three microtubules, and that there are no axial rods. Centrioles are present in the basal bodies of undulipodia, and it is likely that the latter cannot develop without a centriole being present. Typically, a new centriole appears close to an existing one, and at right angles to it. Replication is semi-conservative (Kochanski & Borisy, 1990): when a pair of centrioles replicates, each daughter pair contains one new and one old copy. Maniotis & Schliwa (1991) used microsurgical techniques to produce somatic green mon-

key cells with nuclei but no centrosomes. After a delay of 20–30 hours, a new MTOC capable of nucleating microtubules appeared, but it contained no centrioles, did not partition and so did not produce a mitotic spindle. All this suggests that centrioles may be simple replicators, but apparently this is not universally true, because in a few cases they can arise *de novo*. Thus, mammalian eggs lack a centriole, which is usually supplied by the sperm, but in special circumstances unfertilized mammalian eggs can form centrioles, and then undergo a number of cell divisions.

There are some puzzling features. If centrioles can arise *de novo*, why do they usually arise in the vicinity of pre-existing ones? What is the nature of the information, if any, that is passed from the old to the new structure? It does not follow, because mammalian eggs can develop centrioles *de novo*, that all cells can do so. It is a curious fact that, in contrast to amniotes, no anamniote vertebrate female is known to lay eggs that will develop without fertilization by a sperm, although there are many cases in which the male-derived chromosomes are eliminated. It is tempting to think that a sperm is needed because the egg needs to acquire a centriole.

At mitosis, the centrosome divides, the two daughter centrosomes forming the two poles of the mitotic spindle. Fluorescent antibodies have shown that MTOCs contain some highly conserved proteins. They need not contain centrioles. In higher plants, the microtubules emerge from a region of high electron density that is devoid of centrioles. Centrioles are also absent from the meiotic spindle of mouse oocytes. If MTOCs are isolated from cells and mixed with purified tubulin, they nucleate microtubule assembly: interestingly, the number of microtubules is roughly constant, and equal to the number emerging from the centrosome of the original cell.

The same centriole can serve in turn as the basal body of an undulipodia and as the pole of a spindle. For example, the green alga *Chlamydomonas* has two undulipodia. Prior to division, these are resorbed, and the two basal bodies serve as the poles of the mitotic spindle. Similarly, animal cells cannot simultaneously divide and possess undulipodia. Buss (1987) has discussed the relevance of this for development: we return to this topic in Chapter 15. Presumably, this limitation exists because, depending on the state of the cell—that is, what genes are active and what proteins are present—an MTOC can give rise to an undulipodium, or to a spindle, but not both.

To summarize, it is far from clear whether MTOCs are replicators, and, if so, whether they have hereditary properties. It seems that they are needed to nucleate microtubule formation. Whether an undulipodium or a spindle is formed probably depends on the state of the cell, and not on hereditary properties of the MTOC. The number of microtubules produced, however, does apparently depend on the MTOC, and could be a hereditary property. The characteristic structure of the undulipodium apparently depends on the presence of a similar structure in the centriole. Centrioles typically appear next to pre-existing ones, but need not

do so. It is not clear whether MTOCs carry information needed for the 9 + 2 structure, other than the presence of proteins coded for by nuclear genes, or, if they do, what form the information takes.

8.8 Timing

Remnants of prokaryotes are 3500 million years old. By contrast, the oldest eukaryotic microfossils are not older than 1500 million years. Astonishingly, the time that was needed to pass from inanimate matter to life is four times shorter than that needed for passing from prokaryotes to eukaryotes. Although the origin of eukaryotes involves really difficult steps—such as the elaboration of protein import from the cytoplasm, following the transfer of genes to the nucleus—it is hard to argue that they are more difficult than to establish a genetic code, for example. We offer the following escape from this paradox. During the origin of life, innovations were compared with relatively inefficient competitors. Established prokaryotes are, in contrast, vigorous competitors. Innovations (e.g. loss of the cell wall and origin of the nucleus) are likely to cause disruption and a transient loss in fitness. Thus, special circumstances—in fact, a series of them— must have been needed for these innovations to become fixed, despite the existence of potentially winning competitors (F. Károlyházy, personal communication).

9 The Origin of Sex and the Nature of Species

9.1 Introduction

By sex in eukaryotes, we understand a more-or-less regular succession of meiosis and syngamy. A natural consequence of this is the alternation of haploid and diploid phases in the life cycle.

Eukaryotic sex significantly differs from prokaryotic sex in two crucial respects: the cellular mechanisms are quite different, and the transfer of genetic material in prokaryotes is less frequent and more localized (Maynard Smith *et al.*, 1991). However, there seems to be significant continuity in the molecular mechanisms: sex in either case requires recombination enzymes, many of which are active in repair of damaged DNA as well. Thus, it seems plausible that recombinational repair was a preadaptation for sexual recombination. We mention in passing that there is a theory that selection for the recombinational repair of double-strand DNA damage is responsible for the current maintenance of eukaryotic sex (Bernstein *et al.*, 1981, 1988), but there are severe theoretical as well as factual problems with this theory; we will mention some factual difficulties later.

Although an alternation of haploid and diploid phases follows from sex, a clue to the origin problem may lie in the idea that this alternation existed before the evolution of sexual recombination proper. The first hint that this may have been so comes from the classic paper by Cleveland (1947), where he proposed that the haploid–diploid cycle may have started with a spontaneous diploidization by endomitosis: that is, without syngamy. His suggestions were based on original observations on primitive flagellates (hypermastigotes and polymastigotes). Among them, *Barbulanympha* has a regular endomitosis–meiosis cycle.

Margulis & Sagan (1986) called renewed attention to Cleveland's ideas. In particular, they argued that the alternation of ploidy phases could have a primarily ecological explanation: if the environment alternates in some important factors, this may drive a haploid–diploid cycle, provided the phases are adaptations to different environments. For example, diploids have a smaller relative surface area than haploids, which may confer higher metabolic efficiency. We shall come back to such ideas soon.

9.2 Cellular mechanisms of the haploid–diploid cycle

We focus first on the possible cellular mechanisms connecting the two phases. It is important that some protists have a one-step rather than a two-step meiosis: after syngamy, the two homologous chromosomes become disjunct without premeiotic doubling. One-step meiosis is currently adopted by all archaezoans, and by the Dinozoa (dinoflagellates), Sporozoa (including, perhaps, the malaria parasite) and Parabasalia (trichomonads and hypermastigotes) (Raikov, 1982;

Cavalier-Smith, 1987a). Thus, two-step meiosis with premeiotic doubling is a metakaryotic invention. We do not know how much recombination takes place during one-step meiosis. Because of the lack of premeiotic doubling, chiasmata would not be seen even if recombination occurred: in a two-step meiosis, they are visible only because of the cohesion of sister strands. We will have to wait for genetic evidence to know whether recombination occurs in one-step meiosis. Whether or not recombination takes place, it would be necessary that bivalents be formed, since otherwise aneuploidy would be common.

9.3 Ancient haploid–diploid cycles

Our suggestion is that the stages in the evolution of eukaryotic sex were:

A haploid–diploid cycle, with endomitosis and a one-step meiosis (Fig. 9.1a).

Syngamy and a one-step meiosis (Fig. 9.1b).

Syngamy and a two-step meiosis (Fig. 9.1c).

We now discuss the selective forces that may have been responsible for the various transitions. Before discussing the details, there are two general distinctions that are worth making. The first is between 'mutation' and 'damage' (Bernstein *et al.*, 1988). Mutation alters a DNA molecule into another DNA molecule differing in sequence from the original. Damage alters a DNA molecule into something that is not DNA: it may affect both strands, or only one strand. A mutation, once it has occurred, cannot be recognized as such (except immediately after replication, when the new and old strands may still be distinguishable, as in the processes of proofreading and mismatch repair). In principle, damage can always be recognized, but double-strand damage can be repaired only if a second, homologous, DNA molecule is present to provide the needed information.

The second distinction is between two suggested processes whereby recombination may reduce the deleterious effects of mutation. 'Muller's ratchet' (Muller, 1964) is the gradual accumulation of deleterious mutations in a finite population lacking recombination. The ratchet is arrested by recombination. Although the population must be finite, the process will occur in a large population, provided that the deleterious effect of each mutation is small. A second process was suggested by Kondrashov (1982). In an infinite population, at equilibrium between mutation and selection, the genetic load may be smaller if there is recombination. Although the process does not require that the population be finite, or that each mutation be of slight effect, it does require that the fitness effects of the mutations be synergistic. That is, as the number of mutations in an individual increases, so does the effect on fitness of each additional mutation.

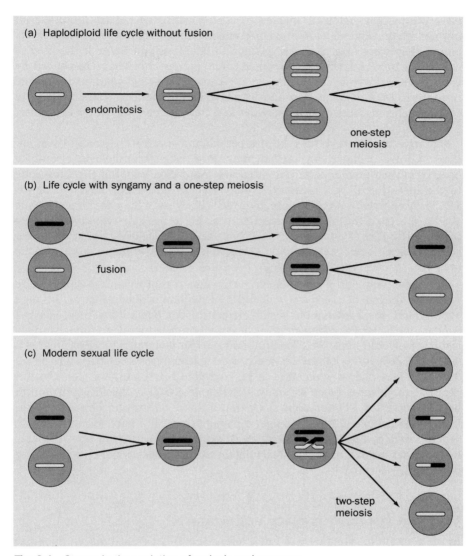

Fig. 9.1 Stages in the evolution of meiosis and syngamy.

Why a haploid–diploid cycle without syngamy?

Szathmáry *et al.* (1990) suggested that a haploid–diploid cycle could have been driven by an environment in which the amount of DNA damage alternated. The potential significance of DNA damage must have been quite high, because the ancient eukaryotes, being archaezoans (lacking mitochondria and peroxisomes),

were presumably not well adapted to coping with free radicals arising from oxygen. They could have lived in environments that were not strictly anaerobic only if they were able to repair damage to double-strand DNA. If the level of oxygen in the environment alternated, archaezoans could have responded by varying their ploidy level accordingly. We know that diploid yeast is much more resistant to DNA damage than haploid yeast, and the latter is more resistant in G2, after chromosome replication, than in G1 of the cell cycle (Kunz & Haynes, 1981).

But why should there have been a periodic reversion to haploidy? In conditions of nutrient limitation, haploids may grow faster than diploids, essentially because of their greater surface area/volume ratio (Box 9A). But there is a snag to the idea that it was nutrient scarcity that favoured haploidy (e.g. Lewis 1985). Archaezoans are not saprophytic cells like yeasts: they are phagotrophic. We have argued that the key invention that led to the eukaryotes was phagotrophy itself. If so, food scarcity could have favoured diploidy, because a larger cell can ingest larger prey: it might also lose less energy in maintaining osmotic pressure.

A possible way out of this difficulty is to suggest that a haploid–diploid cycle preceeded the loss of a rigid cell wall and the evolution of phagotrophy. Although conceivable, the explanation is not attractive. An alternative, more general explanation is that the genetic load at equilibrium between mutation and selection is lower in haploids. (The mutation rate per genome is twice as great in a diploid. If deleterious mutations are not completely recessive, they will be removed from a diploid population in heterozygotes, one mutation per selective death. Thus it takes twice as many selective deaths in a diploid population to balance mutation.) Simulations show that if there is recurrent deleterious mutation, and periods of DNA damage, a population with a haploid–diploid cycle, synchronized so that the diploid phase coincides with periods of DNA damage, has a higher mean fitness than a population that is either permanently diploid or permanently haploid.

Why did syngamy replace endomitosis?

In the beginning, syngamy was not at all necessary for diploidization: the latter could have been achieved by endomitosis (Cleveland, 1947; Hurst & Nurse, 1991). Recently, two yeast mutants have been isolated that become diploid by endomitosis, thus forming complete homozygotes. In the *Saccharomyces cerevisiae* mutant, freshly germinated cells are haploid and mating. They soon become non-mating, and later diploidize. The diploid cells haploidize by normal meiosis (Ono *et al.*, 1990). In the fission yeast *Saccharomyces pombe*, a cell-cycle mutant carries out an extra S phase instead of undergoing mitosis: a simple mechanism for diploidization (Broek *et al.*, 1991). It is interesting that the flagellate *Pyrso-*

BOX 9A Do haploids grow faster?

It is not necessarily true that a diploid grows more slowly than a haploid. In a diploid, not only the material to be synthesized, but also the metabolic machinery, is exactly doubled: hence the division rate would be the same. This is no longer true if cell shape and relative surface matter.

Imagine, for simplicity, that both cell types are spherical. If the diploid cell has twice the volume, it is easy to show that the ratio of surface area to volume of the haploid is 1.25 times that of the diploid. If nutrient uptake, and hence growth rate, are proportional to surface area, the diploid's growth rate is 0.8 that of the haploid. This simple calculation is relatively well supported by experiments. Adams & Hansche (1974) compared isogenic haploid and diploid yeasts in two environments. In high concentrations of nutrients, growth rates were the same: this is not surprising, since in these conditions surface area does not limit metabolism. But when a limiting nutrient was present in low concentration, the diploid fitness was 0.93 that of the haploid.

Weiss *et al.* (1975) analysed the relationship between ploidy, cell volume and metabolism more closely. They studied three situations:

1. If the growth of a yeast population is limited by carbon (e.g. dextrose), haploids and diploids have equal cell volumes, and very nearly equal cell surfaces. Ploidy level is selectively neutral. Among other things, this implies that the time taken to replicate the DNA is not limiting.

2. In a rich nutrient medium, diploids have a volume 1.57 that of haploids, and macromolecular composition is proportional to volume. Again, ploidy is selectively neutral.

3. Complications arise when growth is limited by organic phosphate. Its utilization depends on the surface-bound enzyme, acid phosphatase. The diploid/haploid enzymatic activity, cell-surface area, and cell-volume ratios are 1.44, 1.39 and 1.62, respectively. Thus, phosphatase activity is indeed limited by cell-surface area, and, since diploids are larger than haploids, we would expect the haploids to grow faster. This was borne out by the competition experiments described in the last paragraph.

nympha has an asexual life cycle consisting of several rounds of endomitosis, followed by a series of reductional divisions (Hollande & Carruette-Valentin, 1970). More needs to be learnt about the ecology of endomitotic unicells.

A consequence of the foregoing is that crossing-over would not have been a necessary feature of the early haploid-diploid cycle (Hurst & Nurse, 1991). In fact, there is evidence indicating that chromosome pairing and crossing-over are controlled by two different sets of genes. The same is true for the genetic control of meiotic gene conversion (absolutely necessary for repair of double-strand DNA

damage) and crossing-over (Engebrecht *et al.*, 1990). Synapsis between the homologous chromosomes is mediated by a protein network called the synaptonemal complex. It is believed that recombination nodules between the chromosomes contain the enzymatic machinery of meiotic recombination. Interestingly, there are two kinds: early and late ones. Early nodules are abundant and they are likely to play a role in the initial matching of homologous regions; the late nodules are less abundant and their distribution corresponds with those of chiasmata. The activity of early nodules seems to result in gene conversion only, whereas reciprocal exchange is mediated by late nodules (Smithies & Powers, 1986; Carpenter, 1987).

Since crossing-over would be superfluous in an endomitotically produced diploid, it seems plausible that it originated later, after the haploid–diploid cycle with syngamy, single-step meiosis, and repair by gene conversion had already evolved.

The obvious advantage of replacing endomitosis by syngamy is that arising from hybrid vigour. If deleterious mutations are recessive, or partially so, a diploid arising by the fusion of genetically different haploids will be fitter than one arising by endomitosis. We argue below that the same selective advantage was responsible for the origin of mating types: a cell that fuses with one of a different mating type is more likely to gain the benefits of hybrid vigour.

Although obvious, this explanation must be treated with some caution. The advantage to be expected from syngamy would be much less than that usually arising when crossing inbred lines derived from a normally outcrossing sexual species, because the load of deleterious mutations would be much smaller in an ancestral population with an endomitotic cycle: most seriously deleterious mutations would be eliminated from such a population during the haploid phase. However, although the advantages to be gained from syngamy would be less, they would still be substantial.

In the last section we suggested, albeit rather tentatively, that the advantage of a haploid phase in the life cycle could have been the lower equilibrium genetic load. This depends on the assumption that deleterious mutations are not fully recessive. We are now arguing that syngamy was favoured over endomitosis because most deleterious mutations are partially recessive. There is no contradiction. If, as is likely, most deleterious mutations have some effect in the heterozygote, but a much greater one in homozygotes, then there will be both a higher equilibrium genetic load in diploids than in haploids and an immediate advantage from syngamy.

Why should most deleterious mutations be recessive, or at least more nearly recessive than dominant? This agrees with Wright's (1934) view that, for biochemical reasons, loss mutations would usually be recessive when they first arose, and not with Fisher's (1931) idea that the dominance of most wild-type alleles has evolved by selection for modifiers of dominance. There is biochemical justification for Wright's view. Kacser & Burns (1981) simulated biochemical

pathways, and found that the metabolic flux through a given pathway is usually little altered if the activity of one of the enzymes involved is halved.

Why crossing-over?

This, of course, is the crucial issue, because of its enormous long-term significance. Several population-level advantages of crossing-over are known. We discussed earlier the effect of recombination in reducing the loss of population mean fitness caused by deleterious mutations. Perhaps more important, recombination increases the rate at which a population can adapt to changing conditions, essentially because favourable mutations occurring in different individuals can combine in a single descendent (Fisher, 1930; Muller, 1932). But these are population advantages. Although there is good evidence (from the taxonomic distribution of parthenogens) that they have been important in maintaining sex by eliminating secondarily parthenogenetic populations (Maynard Smith, 1978), we need short-term, individual advantages to account for the origin of recombination.

We need to show that a gene causing crossing-over can be favoured by selection within a population. It turns out that the elimination of deleterious mutations, and the accumulation of favourable ones, can lead to such an individual advantage. Kondrashov (1988) found that synergistically acting deleterious mutations can select for recombination within a population.

The experimental evidence that deleterious mutations act synergistically is limited. It would therefore be useful to have some theoretical expectation. In general, it is impossible to predict the nature of epistasis, but it is possible for those mutations that affect metabolism by reducing enzymatic efficiency. Using the metabolic control theory of Kacser & Burns (1973, 1979), Szathmáry (1993b) was able to deduce the nature of epistasis between enzyme mutations.

Metabolic control theory calculates the flux of metabolites through a pathway, as a function of the activities of enzymes catalysing the individual steps of that pathway. It is easy to calculate the reduction in flux caused by a diminished enzymatic activity. Whether mutations have a synergistic or antagonistic effect on fitness depends on whether selection favours an optimal or a maximal flux, as follows:

- If selection favours a constant, optimal concentration of some metabolic intermediate, the effects of mutation on fitness are synergistic, so that epistasis will be favourable for sex, as required by Kondrashov's theory.

- If selection favours the most rapid synthesis of the end product of the pathway, the effects of mutation are antagonistic, and recombination will be disadvantageous.

These theoretical results fit rather well with Kondrashov's mutational theory. Unicellular organisms, and prokaryotes in particular, are often limited by the supply of nutrients, so that the flux through some pathway limits their metabolism, and hence their division rate. It is therefore consistent with the theory that prokaryotes lack meiotic sex, and many protists lack sex. A second prediction, which partly overlaps the first, is that *r*-and *K*-selected organisms should have little and much recombination, respectively—a prediction that is also born out (Bell, 1982).

Maynard Smith (1979b, 1988b) showed that, if there is directional selection for a trait determined by genes at several loci, there will also be selection for a gene increasing the rate of recombination between those loci: however, if selection on the trait is normalizing, there is selection for reduced recombination. Hamilton (1980) argued that an arms race between host organisms and their parasites is a likely cause of selection for change, and Bell & Maynard Smith (1987) found that such a system can select for increased recombination. Thus, selection on synergistically acting deleterious mutations, and directional selection on polygenic traits, can favour the evolution of crossing-over. Either, or both, processes could have been responsible for the origin of crossing-over in meiosis. An important point is that this step, although it had immense consequences, was not a big one: both a one-step meiosis with pairing of homologous chromosomes, and the enzymes needed for cutting and rejoining chromosomes, were already present.

Why a two-step meiosis?

Crossing-over does not require a two-step meiosis. Why, then, is there premeiotic doubling? At first sight, the process seems absurd: the function of meiosis is to halve the number of chromosomes, but the first thing that happens is that the number is doubled. The first explanation that we know of comes from Bernstein *et al.* (1988). They argued that there are dozens of specific enzymes eliminating various types of single-strand DNA damage, yet it is impossible to have a specific enzyme for every type of damage, since the number is countless. It is, however, straightforward to convert, by replication, a DNA with a damaged single strand to two DNAs, one perfect and one with a double-strand lesion. Thus, the function of premeiotic doubling is to convert irreparable single-strand damage to reparable double-strand damage. The hypothesis is ingenious. The fact that two-step meiosis originated with metakaryotes is in favour of it: their predominantly aerobic metabolism, with mitochondria and peroxisomes, calls for efficient DNA repair because of the formation of oxidizing free radicals.

Although ingenious, the idea has its difficulties. The repair of single-strand damage occurs in clonal prokaryotes and in ordinary mitosis. Premeiotic doubling only extends to meiosis a process of chromosome replication that preceeds every mitotic division. If its function is to facilitate single-strand repair, it is needed only to repair the damage that has occurred since the last mitosis.

Haig & Grafen (1991) suggested that a possible advantage of a two-step meiosis is that a so-called *sister-killer* allele could not spread. Imagine an allele able to kill its sister gamete. With a one-step meiosis, such an allele would spread when rare, since it will find itself in a heterozygote prior to division. This is not so with a two-step meiosis, because if the first division is reductional at a locus the second will be equational, and vice versa (the first division will be reductional if there was no crossover between the locus and the centromere). The *sister-killer* might be active in either the first or second division, but since it is uncertain which division will be reductional, there is a 50 per cent chance that it will kill itself. Hence, there is no net reproductive advantage to the allele.

Haig (1993) suggested that some peculiar division mechanisms in microsporidia (belonging to the Archaezoa) and rhodophytes may be adaptations that resolve the problem in an alternative way (see Box 9B).

BOX 9B Some alternatives to a two-step meiosis

As explained in the main text, Haig & Grafen (1991) suggested that one function of a two-step meiosis might be to prevent the spread of sister-killer mutants. Haig (1993) has interpreted some peculiar division mechanisms as alternative solutions to the same problem.

Microsporidia have been known to undergo a single-step meiosis, and until recently there was no evidence for crossing-over or a synaptonemal complex (Raikov, 1982). However, Canning (1988) has recently discovered that a synaptonemal complex is formed after a diploid meiosis, which follows syngamy: a second round of syngamy then follows between the diploid products of mitosis (Fig. 9.2). Due to the lack of crossing-over, the two synaptonemal complexes each consist of a pair of non-sister chromatids. If it is assumed that the chromatids of the two 'bivalents' assort independently in the first division, then either the first or second reduction division contains non-sister homologues. This results in the same type of uncertainty for a sister-killer as ordinary meiosis with crossing-over.

An alternative strategy is suggested for some red algae, notably *Choreocolax* and *Polysiphonia* (Fig. 9.3). Regular cycles of endomitosis (DNA replication without cell division) and depolyploidization (division without DNA replication) occur. Ploidy may be as high as 128. Presumably, some kind of synaptonemal complex must exist to prevent non-disjunction of homologues in division, although there is no cytological evidence. For example, at a ploidy level of 16, eight bivalents must form, not necessarily consisting of one maternal and one paternal chromosome. If chromosomes assort independently, a binomial distribution of alleles will arise at the diploid level. Hence, as the initial high ploidy level increases, the probability that a sister-killer would attack itself approaches one-half.

These peculiar division mechanisms do at least make sense as adaptations preventing the spread of sister-killers. They do, therefore, support the idea that the problem is a real one, and so make it more plausible that a two-step meiosis may initially have had a similar function.

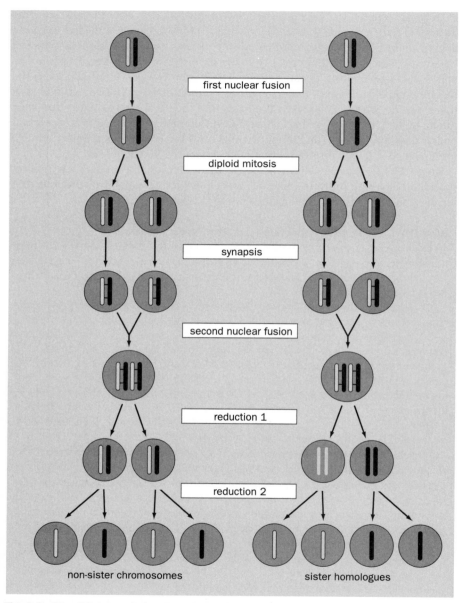

first nuclear fusion

diploid mitosis

synapsis

second nuclear fusion

reduction 1

reduction 2

non-sister chromosomes

sister homologues

Fig. 9.2 The division cycle in microsporidia (after Haig, 1993).

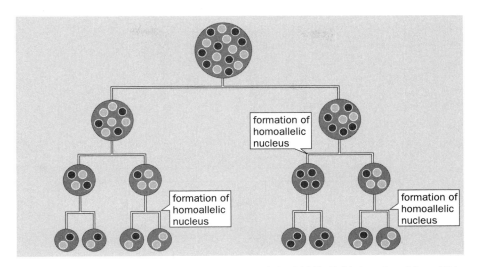

Fig. 9.3 The reduction of ploidy in red algae (after Haig, 1993). Ploidy is reduced from 16 to 2 in three divisions. Open and filled circles represent different alleles at a locus.

Conclusions

It is possible to propose selective explanations for the transitions illustrated in Fig. 9.1, but these are unavoidably tentative. Two problems worry us in particular. First, in explaining a haploid-diploid cycle without syngamy, the role of diploidy in making possible double-strand repair is clear, but the advantages of a reversion to haploidy are more problematic. It is important to learn more about the ecology of existing protists with a comparable life cycle. Second, the repair function of premeiotic chromosome doubling is not wholly convincing, because it is relevant only to the repair of single-strand damage occurring since the last mitosis. We need to know more about existing organisms with a one-step meiosis, and in particular whether genetic crossing-over occurs in them.

9.4 Mating types and the origin of anisogamy

Even in sexual organisms in which there is no morphological difference between the gametes—so-called isogamous organisms—it is a general rule that there are two mating types, + and − : gametes of each type fuse only with gametes of the other. This state of affairs probably evolved very early from one in which any gamete could fuse with any other. Hoekstra (1982, 1987) has suggested a possible mechanism (Fig. 9.4). In the initial, promiscuous, phase two separate genes, A and B, coded for two proteins, corresponding to what an engineer would, appropriately, call male and female fittings. This can readily be converted

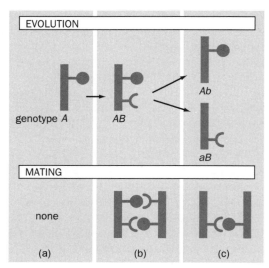

Fig. 9.4 The origin of + and − mating types (after Hoekstra, 1987). (a) Gene *A* specifies a surface protein not concerned with cell fusion. (b) Gene *B* specifies a protein causing cell fusion: any cell will fuse with any other. (c) Loss mutations, *A* → *a* and *B* → *b*, produce two mating types, each of which will fuse only with the other (or with the ancestral *AB*, while that still exists). At this stage there will be selection for tight linkage between loci *A* and *B*, because recombinant *ab* genotypes cannot fuse. During the initial spread of *Ab* and *aB*, there would have to be a selective advantage in not fusing with one's own type, probably through the covering up of deleterious mutations: there would also be selection maintaining approximate equality of frequency between the two types.

into a system with two mating types by loss mutations, *A* → *a*, and *B* → *b*: the two mating types would then be *Ab* and *aB*. Because recombination between these would produce *ab* genotypes, which could not fuse at all, there would be selection for tight linkage between the two loci. In fact, in the green alga *Chlamydomonas reinhardii*, the mating-type 'locus' consists of a cluster of tightly linked genes, all functionally involved in the mating process.

The model requires the initial presence of two genes, both of which are needed for fusion. This, however, need not be a difficulty, since one of the loci might originally have coded for a surface protein with some other function. Hoekstra (1987, 1990) sees great difficulty in accounting selectively for the change from *AB* to *Ab* and *aB*, because he believes it to be associated with a twofold disadvantage. Thus, in a population consisting of equal numbers of *Ab* and *aB*, an *AB* mutant could fuse with any other gamete, whereas *Ab* and *aB* could fuse with only half of the other gametes. It is not obvious, however, that this would confer any great selective advantage on *AB*. After all, a gamete that cannot fuse with the first partner it encounters can always try again. Provided it is selectively advantageous for a gamete to fuse with one different from itself, it is rather easy

to see how an all-*AB* population could be invaded by *Ab* and *aB* mutants, and how the latter would increase in frequency together until *AB* was eliminated.

There remains the question why it should be an advantage, first to fuse at all and then to fuse with a genetically different partner. This question was discussed in section 9.3. We think that the most plausible answer is that fusion was preceded by endomitosis in a haploid–diploid life cycle: if so, fusion would have the advantage of covering up deleterious mutations.

The next stage in the evolution of gamete differentiation is the transmission of organelles by only one parent. Cosmides & Tooby (1981) were the first to suggest that conflict between the organelles from the two parents may have caused the evolution of anisogamy. They argued that genes in, for example, mitochondria that caused an increase in gamete size would be favoured by selection. More recently, Hurst & Hamilton (1992) have argued that, since conflict between organelles would be damaging to the zygote, there would be selection for nuclear genes able to prevent this conflict by ensuring uniparental inheritance: such inheritance is, in fact, common, even in isogamous protists (for reviews, see Grun, 1976; Whatley, 1982). In *Chlamydomonas*, for example, mitochondria are inherited only from the − mating type and chloroplasts only from the + type. There is evidence that this uniparental inheritance is controlled by nuclear genes.

Clearly, uniparental inheritance of organelles can evolve more easily if there are just two mating types. It would be difficult if there were multiple incompatibility types, each of which could mate with any other type but not its own. In support of their view that there is a causal connection between uniparental inheritance, conflict between organelles and the existence of two mating types, Hurst & Hamilton point to the fact that a multiple incompatibility system is found only in those groups in which there is no cytoplasmic fusion at mating, and hence no possibility of conflict between organelles: that is, in the basidiomycetes and among ciliates. In the latter group, there is exchange of micronuclei at conjugation, but no cytoplasmic mixing, and there are multiple mating types. It is particularly persuasive that, in hypotrichs with both conjugation and gamete fusion, there are multiple incompatibility types determining conjugation, but only two mating types determining gamete fusion.

A system of multiple mating types is unlikely to be primitive, if only because it is biochemically more difficult to achieve than either the bipolar or unipolar systems illustrated in Fig. 9.4. However, if there were no disadvantage to having more than two types, it is likely that multiple types would evolve. A new rare type, able to mate with all other members of the population, would have an advantage over the commoner types, even if the advantage was less then the twofold one suggested by Hoekstra. The selective process is the same as that, first suggested by Wright (1939), maintaining multiple incompatibility types in higher plants. Hurst & Hamilton (1992), therefore, are almost certainly right to argue that two mating types are usually retained because uniparental in-

heritance of organelles would not work with more that two types, and that, when there is no cytoplasmic fusion, this constraint is removed, and multiple types can evolve.

In what circumstances would a nuclear gene, R, destroying the mitochondria in its own gamete before or immediately after fusion, be established in a population? It could not spread in a promiscuous population, because $R \times R$ matings would be lethal. But if there already existed a locus determining mating type, with two alleles, then the gene R could spread, provided that it was tightly linked in coupling to one of the alleles at the mating-type locus.

The next stage is the evolution of size dimorphism of gametes. Parker *et al.* (1972) proposed a model in which a population of 'male' individuals producing small motile gametes could be invaded by 'females' producing large non-motile ones. The essential assumption is that the probability $P(S)$ that a zygote will survive to reproduce is an increasing function of zygote size, S. The model was further developed by Maynard Smith (1978), who showed what form the $P(S)$ curve must have if females are to invade an all-male population, giving rise to a stable bisexual population. Parker *et al.* did not assume that there were initially two mating types, but their model works well with that assumption. Knowlton (1974) provided confirmation of this idea in a comparative study of gamete dimorphism in a group of colonial algae, the Volvocales. Species with small colonies have equal-sized motile gametes; species with colonies of intermediate size produce motile gametes of variable size; species with the largest colonies produce large, non-motile gametes and small, motile ones.

To summarize, we suggest the following stages:

- A promiscuous population, in which fusion replaced endomitosis as a means of achieving diploidy in a haploid–diploid life cycle.
- The evolution of two mating types, ensuring that the advantage of 'hybrid vigour' resulted from fusion.
- Uniparental inheritance of organelles, under the control of nuclear genes, preventing interorganelle conflict.
- Size differentiation between the gametes produced by the two already existing mating types, as a direct consequence of the relation between zygote size and chances of survival.

We regard this scenario as plausible, but not as the only possible one. To mention some interesting alternatives: Hickey & Rose (1988) suggested that fusion was initially under the control of genes in an endosymbiont, perhaps a virus, ensuring its own horizontal transfer, as conjugation in *E. coli* can be caused by the F plasmid; Hoekstra (1990), also postulating horizontal transfer of a symbiont as the driving force, suggested that cell fusion may have been asymmetric from the outset; and Hurst (1990) argued that anisogamy may have evolved because it reduces parasite diversity.

9.5 Sex and the nature of species

At least in higher animals, in a single geographical region, variation is not continuous. A competent bird-watcher will identify the birds he sees as blue tits, great tits, coal tits, marsh tits, and so on, and other observers will agree with his identifications. Why should this be so? Three different hypotheses have been suggested:

- There are species for the same kind of reason that there are chemical elements. Species represent stable states of living matter, determined by (as yet unrecognized) 'laws of form' (Bateson, 1894; Webster, 1984). The facts about variation discussed below seem to rule out such an explanation.

- Species exist because they are adapted to discontinuous ecological niches. Sometimes this is correct: for example, herbivorous insects may be adapted to different food plants, and parasites to different hosts. Clearly, this cannot be a general explanation, because it explains species in one taxon by saying that there are species in another: it is unavoidably circular.

- Species exist because of sexual reproduction. Although this is the generally accepted view, it is perhaps not quite as obviously true, either in theory or empirically, as one could wish.

Theory

Roughgarden (1979) considered the following model. There is a resource axis, and an array of clones, each one best adapted to a certain point on the axis. There is a constant rate of resource supply at each point of the axis. Each clone has a resource utilization curve, having its maximum at the point of best adaptation. The question of interest is the densest stable equilibrium packing of clones on the axis. It turns out that there is no such packing: a novel type of clone can always invade when rare, and in the limit there is an infinite number of clones, each with zero density.

The situation changes dramatically, as shown by Hopf & Hopf (1985), when there is a reduction in per capita growth rate with decreasing density. The authors had sexual reproduction in mind: the reduction occurs because of the cost of finding a mate. There is always a species density that cannot be increased further: novel species cannot invade when rare (see Szathmáry, 1991a, for a review). This result is consonant with the existence of discrete species. Note, however, that this analysis does not explain the existence of species: it assumes that species already exist. It merely shows that one cannot have an infinite number of reproductively isolated species. But it does not rule out the following possibility, which we will call the 'sexual continuum' model. Imagine that two populations can interbreed if they are not too far apart phenotypically. Thus, even if individuals from 'distant' populations cannot interbreed, there are no

proper species because, after a sufficient number of generations, a gene from a population at one phenotypic extreme can pass to a population at the other extreme. We return to this possibility after discussing the empirical data.

Observations

There are organisms that reproduce parthenogenetically. We would predict that, in such groups, discontinuities between species should not exist. The facts are not quite so simple: we review them separately for plants, animals and bacteria.

Parthenogenesis in plants

Parthenogenesis is common in plants, but in most such groups (e.g. *Rubus*, *Hieraceum* and *Taraxacum*) there are also sexual forms, and new parthenogenetic clones arise repeatedly from the sexuals. There are, however, some groups in which sexual forms are rare, and probably contribute little to the pattern of variation. For example, the lady's mantles, *Alchemilla*, of Europe are almost all parthenogens (although there is a dwarf sexual alpine species, *A. pentaphylla*; Briggs & Walters, 1984). There are some 300 morphologically distinguishable forms. Although many of these are widely distributed, and may represent single clones, the large number of forms, and the trivial nature of the differences between them, distinguishes *Alchemilla* from most sexual taxa.

It is true that the number of clones of *Alchemilla* is not infinite, but this is to be expected in a world that is finite, and not infinite as in Roughgarden's model. The *Alchemilla* data, then, support the view that distinct species exist because of sexual reproduction. A more serious difficulty is that there are also sexual groups which are not easily divided into species. For example, *Potentilla glandulosa* exists in a number of ecologically adapted populations from the coastal hills of California to the Sierra Nevada (Clausen & Hiesey, 1958). Although the extreme forms could not survive in the environment of the other, there are no barriers to gene flow: the whole complex forms a sexual continuum. A more extreme example is afforded by the white oaks of North America (Muller, 1951; Hardin, 1975). There are some 30 named 'species', but all are known to hybridize in nature with at least some of the others, and the whole set form a single sexual continuum. Such a pattern of extensive variation without reproductive isolation is probably characteristic of groups that have undergone a recent adaptive radiation. Thus, it is found in plants that have recently colonized oceanic islands: for example, the genus *Metrosideros* provides the main canopy trees of the wet highland forests of Hawaii, and also the first colonists of recent lava flows, yet there seems to be no barrier to gene flow.

The pattern of variation in plants corresponds to that expected if species are a consequence of sex, but with two reservations, First, the number of clones of asexual taxa, although large, is not infinite: this should not surprise us in a finite world. Second, some sexual taxa are not readily divided into species: these are probably taxa that have recently radiated but not yet speciated.

Parthenogenesis in animals

Most parthenogenetic animals are of relatively recent origin, and arose by hybridization between sexual species. In so far as they fall into distinct 'species', this is because their sexual ancestors did so. It is sometimes clear that a single asexual 'species' consists of several clones that arose independently by hybridization between the same two sexual species. Few animal taxa appear to be wholly parthenogenetic. Perhaps the most important fact to emphasize is that of geographical variation. Populations from the extremes of the range of a single species, although connected by a series of intermediate forms, may differ from one another as much as do many pairs of good sympatric species: it is a matter of convenience and convention that they are placed in the same species. Such cases are sufficient to disprove any 'laws of form' hypothesis about species.

The conventional view, then, is that in animals, and to a lesser degree in plants, the existence of discrete species is a consequence of sexual reproduction. There are, however, few studies in which the pattern of variation in related sexual and asexual taxa have been compared with the explicit aim of testing this hypothesis. Holman (1987) compared previous taxonomic studies of bdelloid (asexual) and monogonont (sexual) rotifers. Surprisingly, he found that synonymous species names are less common in the bdelloids than in the monogononts, suggesting that species are easier to recognize in the former group. It is hard to interpret this finding, because it depends on the habits of taxonomists as well as on the real pattern of variation in nature. But Holman's conclusion is potentially important: if it could be confirmed by a direct morphometric study of the two groups, it might lead to a profound change in our views on the nature of species.

Parthenogenesis in bacteria

Although bacteria do not have meiosis and syngamy, they exchange chromosomal genes by a variety of means. It is customary to give them specific names— *Escherichia coli*, *Salmonella typhimurium*, *Neisseria meningitidis*, and so on. Is this mere convenience, or do bacteria really belong to reproductively isolated groups? Even if they do, it is clear that the names used at present do not correspond to monophyletic groups. For example, *Shigella* is distinguished from *Escherichia* because it causes unpleasant symptoms in humans, but a molecular analysis shows that *Shigella* should be included within *Escherichia* (Ochman *et al.*, 1983). It does not, of course, follow that there are no reproductively isolated groups—it may just be that we have wrongly identified them. However, even if we confine ourselves to the transfer of chromosomal genes by mechanisms involving homologous recombination (for a review of the evolutionary consequences of such processes, see Maynard Smith *et al.*, 1991), it is not clear that species exist. If homologous recombination is to occur, nucleotide divergence must not be too great: perhaps 20 per cent divergence is close to the upper limit. But there is no reason why two bacteria differing by much more than this should not form part

of a sexual continuum via intermediates. In one relevant study, Cohan *et al.* (1991) measured the frequency of transformation between natural isolates of *Bacillus*. He found wide variation. The main determinants were sequence similarity and restriction-modification type.

Conclusions

As always, the facts are messy. However, it does seem that a clear differentiation into species is found only, but not quite always, in organisms with meiotic sex, living in the same place.

Fortunately, we do have a rather satisfactory theory about how reproductive isolation arises. Although it is possible to construct genetic models of how isolation could arise sympatrically (e.g. Maynard Smith, 1966), it seems that most speciation events required a period of spatial isolation (Mayr, 1942). Populations diverge genetically when they are spatially isolated. If they subsequently meet, there are three possibilities: (1) they are still sufficiently similar to merge into a single population; (2) the divergence is so great that hybridization is impossible, or leads to sterile hybrids; or (3) hybrids are formed, but are of lowered viability or fertility. If inviable, infertile or sterile hybrids are formed, there will be selection for traits that prevent hybridization (Dobzhansky, 1951; see Coyne & Orr, 1989, for evidence that such selection has been effective in *Drosophila*).

In cases (1) and (2), two species arise, where there was only one before. It is, of course, possible that one of the new species will be competitively eliminated. However, we do have a mechanism that will, over a sufficient period of time, generate an increasing number of reproductively isolated species. Why, then, does Roughgarden's argument not apply: why does the number of species not increase indefinitely? There are two possible answers:

- In a finite world, rare species will be eliminated by chance. If this were the only reason, we would expect sexual species to be as tightly packed as parthenogenetic clones: the *Alchemilla* pattern would be found also in sexuals.
- There is a 'cost of rarity' for sexuals arising from the difficulty of finding a mate, as suggested by Hopf & Hopf (1985).

We conclude that, despite the intense interest of biologists in the 'species problem' for over two centuries, the answers are not as clear as they should be. Observationally, there is little to support the idea that species, like chemical elements, represent possible stable states of matter: the facts of geographical variation are sufficient to refute this. Most data on animals and plants are consistent with the idea that the existence of discrete species is a consequence of sexual reproduction, but there are difficulties. Some sexual taxa do not fall neatly into species, probably because they are recent radiations, and reproductive isolating mechanisms have not had time to evolve. More important, there have been few explicit attempts to test the idea by comparing sexual and asexual taxa:

one of the few such investigations, on rotifers, may contradict the conventional view.

On the theoretical side, it is plausible that a 'sexual continuum', in which there are no discrete species and individuals can mate with others not too distant from themselves, would break up into species because of the lowered viability and fertility of hybrids between more distant parents. However, we are not aware of any explicit model demonstrating the instability of a sexual continuum. In any case, it is important to distinguish between processes leading to the existence of discrete sexual species, and processes that prevent the numbers of such species increasing without limit, or at least increasing up to the same limit as clones within an asexual taxon.

10 Intragenomic Conflict

10.1 Introduction

The idea that there may be a conflict between different genetic elements—for example, between chromosomal and mitochondrial genes—is a natural one. A wide range of phenomena, from distorted sex ratios and male sterility in plants to the evolution of chromosome form, may best be understood in terms of such conflict. Before discussing examples, however, it is useful to define 'genomic conflict'. Suppose that two genetic elements, *A* and *B*, influence some phenotypic trait, T. Suppose, further, that some value of the trait, T_A, would evolve if only genetic elements *A* were effective, and an alternative value, T_B, if only elements *B* were effective. If T_A and T_B are not the same, then there is a potentiality for genomic conflict between elements *A* and *B*. In this chapter, we are mainly concerned with cases in which the two genetic elements are present in the same cell: hence the term 'intragenomic' conflict. However, the definition can be applied when the elements are in different cells in the same organism, or in different organisms: 'parent–offspring conflict' (Trivers, 1974; Parker & MacNair, 1978) is a case in which there is a conflict between genes expressed in different organisms.

The occurrence of intragenomic conflict depends on how the genes are transmitted. Imagine a single-celled asexual haploid organism, with a single chromosome containing no transposable elements, one copy of which is transmitted to each daughter cell. Suppose further that the cell contains no symbionts or accessory genetic elements. There would then be no possibility of intragenomic conflict. In practice, conflict arises because cells contain transposons, plasmids, organelles and symbionts, and are sexual.

Most biologists find the phenomena of intragenomic conflict unfamiliar and unexpected. It might seem that a good way of making sense of the subject would be to classify the genes concerned according to whether they are in parasites, mutualists, organelles or chromosomes. The snag is that parasites can evolve into mutualists and vice versa, symbionts can become organelles, genes can transfer from organelles to chromosomes, and chromosomal genes may be well behaved, transposable or 'driving'. For these reasons, we will discuss intragenomic conflict under four heads, according to the different ways in which genes can obtain an unfair advantage over their fellows.

Meiotic drive. Chromosomal genes can distort segregation, and so gain a disproportionate representation among gametes. This topic is discussed in section 10.2.

Transposition. A second way in which genes can obtain an unfair advantage is by replicative transposition. In section 10.3, we describe how transposition is regulated. It turns out that, at least in some cases, transposition is regulated by the transposon itself. Although at first sight surprising, this makes sense if one remembers that the future of a transposon depends on copies of it being

transmitted to offspring of the individual in which it now is. Transposons have been very successful. If one counts success in terms of amount of DNA (rather than amount of protein coded for), transposons are more successful than their eukaryotic hosts. In section 10.3, we describe some effects of this success on the evolution of chromosomes and of organisms.

Conflict between organelles. Since there is no mechanism to ensure the exact segregation of organelles, as there is for chromosomal genes, any gene giving an organelle an advantage relative to others is likely to spread, even if it lowers organism fitness. Section 10.4 is concerned with how this potentially damaging conflict between organelles is controlled.

Conflict between genes in organelles and in the nucleus. The mode of transmission of genes in organelles and in the nucleus is different: in particular, organelles are often inherited uniparentally. This leads to a conflict of interest between nuclear and cytoplasmic genes. Some effects of this conflict, particularly on the sex ratio, are discussed in section 10.5.

10.2 A fair meiosis

Meiotic drive

A diploid zygote carries two potentially different alleles at every locus. In a fair meiosis, the two alleles have exactly equal probabilities of being transmitted to the next generation. Hamilton (1967) pointed out that there will be strong selection in favour of any allele with 'meiotic drive': that is, with a better-than-evens chance of transmission. He further showed that if the 'driving' gene is on a sex chromosome, its spread can lead to the extinction of the population. In practice, however, meiosis is usually fair. There are two main reasons for this:

- There may be rather few things a gene, or linked group of genes, can do to increase its chance of transmission.
- Meiotic drive will usually have adverse effects on the transmission of genes at other loci, and there will therefore be selection for modifiers of meiotic drive. In Leigh's (1971) expressive phrase, the 'parliament of genes' will oppose meiotic drive.

Some examples will show that drive is not just a hypothetical possibility. In mice, males heterozygous for a *t* allele (now known to be a chromosomal region of some 2×10^4 kb, within which recombination is suppressed by inversions; Silver, 1985) produce offspring almost all of which carry *t*, and not its wild-type allele. Sperm with the + allele appear normal in morphology and locomotion, but, for unknown reasons, are unsuccessful in fertilizing eggs. Presumably, during spermatogenesis, the *t* allele causes some damage to + sperm. *t* alleles do not go to fixation in mouse populations, because *t/t* males are sterile. As a second

example, *Drosophila melanogaster* males heterozygous for the gene *SD* (segregation distorter) transmit the gene to 95 per cent or more of their progeny (Hartl & Hiraizumi, 1976). Sperm not carrying *SD* are non-functional. We return to this example later. As a final example, in the mosquito *Aedes aegypti*, males carrying a driving Y chromosome produce progenies with from 6 to 43 per cent of females, depending on the origin of the X chromosome (Wood, 1976).

These examples are typical in that all concern male meiosis in species with internal fertilization. To understand this, we review briefly the circumstances in which meiotic drive would be favoured at the gene level.

Isogamous species

Consider a gene in a diploid gametocyte that was able to destroy, or render inactive, gametes carrying its wild-type allele. When rare, such a gene would have little selective advantage: its chances of fusing with a gamete of opposite mating type would not be greatly increased by killing its 'sibling' gametes. Genes at other loci, however, would have much to gain by suppressing its effects. It is, therefore, not surprising that such genes have not been reported.

It seems, then, that the only type of 'meiotic drive' gene that would be strongly favoured in an isogamous species is one that could convert its allele to resemble itself: that is, one that caused 'biased gene conversion'. However, if there is a bias in gene conversion, this will be because the enzymes replicating DNA make some types of conversion more frequently than others. This could be selected for. Bengtsson (1985) has made the ingenious suggestion that, if mutation is biased, then, since most mutations are harmful, selection would favour replicases which, faced by a heterozygote, make conversions in the direction opposite to that of mutation: for example, if the deletion of a base is a commoner mutation than the addition of a base, selection would favour a replicase which, faced by a mismatch during meiosis, added rather than subtracted a base. For whatever reason, biased conversion does happen, and results in non-Mendelian ratios, but it probably arises from selection operating on the replication machinery, and not on the genes that are converted. Once bias has evolved, however, it will have a directional effect on the kinds of mutations that occur.

Anisogamous species

Females. A gene which could ensure that it moved to the egg pronucleus, and that its allele passed to a polar body, would be strongly favoured. Surprisingly, rather few such cases have been reported. Some females of the butterfly *Acraea encedon* produce all-female broods (Owen, 1970). The trait is inherited from mother to daughter. Since, in butterflies, the female is the heterogametic sex, these facts are consistent with the presence of a driving gene on the Y chromosome. A similar but more complex situation has been described in *Danaus chrysippus* (Smith, 1975). It may be that the rarity of reported cases of meiotic drive in females arises because drive is usually noticed through its effect on the sex ratio, so that drive in females will be noticed only in those groups in which

the female is the heterogametic sex. It is, however, striking that no cases have been reported in *Drosophila*. In mammals, Kaufman (1972) reported that, in XO mice, chromosome sets lacking an X pass preferentially to the polar body, but this can hardly be regarded as an example of a meiotic drive gene that has been favoured by selection.

These data suggest that genes in the meiotic chromosomes of eggs can do rather little to influence their chances of getting into the egg pronucleus. It is not clear how far this helplessness is a consequence of selection, at the organism level, for suppression of meiotic drive, and how far it is an unselected consequence of chromosomal condensation: condensation probably evolved because condensed chromosomes are easier to move about, but has the secondary consequence that genes are inactivated.

Males. From a selective viewpoint, males of anisogamous species differ from isogamous individuals in two ways. First, the gametes of different males compete intensely for fertilization. Second, in species with internal fertilization (or if breeding occurs in pairs or small groups) the gametes of a single individual compete with one another. Therefore, a gene that can destroy gametes carrying its allele will be favoured. This will be true even when the gene is rare, provided there is competition between the sperm of a single male. In fact, most observed cases of meiotic drive occur in this context.

In those cases that have been studied in detail, it turns out that at least two loci are needed to generate meiotic drive. Perhaps the best-studied case is segregation distorter, *SD*, in *Drosophila* (for the population genetics of this case, see Charlesworth & Hartl, 1978, and references therein). There are two tightly linked loci: the distorter locus with alleles *SD* and $+$, and a 'responder' locus with alleles *Rsp* and *Rsp*$^+$. The *SD* product inactivates any gamete with the allele *Rsp*$^+$, but not with allele *Rsp*. It follows that the original *SD* mutation could invade the population only if it arose in coupling with *Rsp*: an *SD* gene in coupling with *Rsp*$^+$ would destroy itself.

To generate meiotic drive, then, both a gene producing a 'toxin' (*SD*), and a 'target' gene (*Rsp*), are needed. In the case of an autosome, two further conditions are required:

- The loci must be tightly linked: otherwise, the necessary linkage disequilibrium, with *SD* and *Rsp* in coupling, will be destroyed.

- The *Rsp* gene must initially be rare. If, before the *SD* mutation occurred, most chromosomes carried *Rsp*, *SD* would have little advantage.

Hurst & Pomiankowski (1991a) pointed out that, because of these conditions, autosomal meiotic drive may arise rarely, but that meiotic drive affecting the sex chromosomes can be established more easily, essentially because of the absence of recombination between the X and Y chromosomes. Thus, on an autosome an *SD*-type gene can spread only if it occurs close to a pre-existing, but rare, *Rsp*-

type target gene. In contrast, if there is any consistent allelic difference between the X and Y chromosomes that could act as a target, then an *SD*-type mutation occurring anywhere on the Y chromosome, attacking the target on the X, will spread, provided there is no recombination. The same is true of an *SD*-type mutation on the X attacking a target on the Y.

Once established, meiotic-drive genes are likely to lower the chances of transmission of other genes present in the same cells, both because of their direct adverse effects on viability and fertility, and, for sex-linked drivers, through their effect on the sex ratio. Therefore, modifiers that suppress meiotic drive will spread under selection (Eshel, 1985). Crow (1991) argued that the reason why meiosis is usually so exact is that, whenever segregation distorters have arisen, their effects have been suppressed by modifiers elsewhere in the genome. In so doing, he is extending Leigh's appeal to the parliament of the genes. While accepting this argument, Hurst & Pomiankowski (1991b) surmised that 'prevention is better than cure': that is, that there may be general features of meiosis that evolved because they make it difficult for meiotic drive to arise. What might such features be? Hamilton (1967) suggested that the inactivation of the Y chromosome is such a feature. Another may be the fact that, in general, the phenotype of the sperm depends on the paternal genotype, and not on its own. Haig & Grafen (1991) suggested that recombination may have evolved because, by destroying linkage disequilibrium, it prevents the spread of driving genes: one snag with this suggestion is that, if recombination was not favoured for other reasons, organisms would not indulge in sex in the first place. Finally, according to Hurst & Pomiankowski (1991b) there may be processes that ensure that, if drive does occur, all the gametes produced are destroyed.

Summarizing, we are left with the feeling that the avoidance of intragenomic conflict in meiosis may be easier to explain than it seems at first sight. Selection for meiotic drive in isogamous organisms, and in males with mass spawning, will be weak. In females, despite the strong selection in favour of alleles with preferential access to the egg pronucleus, meiotic drive may be made difficult by features (e.g. chromosome condensation) that evolved for other reasons, although it is intriguing that the first chromosomes to undergo condensation during meiosis are often the sex chromosomes—that is, the chromosomes most likely to undergo meiotic drive. The problem arises mainly in males with internal fertilization, or mating in small groups. In these, several cases of segregation distortion are known. It is easy to see why specific suppressors of particular driving genes should evolve. A more difficult question is whether there are general features of meiosis that evolved because of their effect in suppressing meiotic drive.

B chromosomes

This is a convenient point at which to discuss the biology of B chromosomes, which have evolved several ways of achieving unfair access to future genera-

tions, including distortion of meiosis. B chromosomes are small, often hetero-chromatic, supernumerary chromosomes. They vary in number between populations, between individuals, and between nuclei within an individual. They have been described in some 1000 species, and are probably present in 10–15 per cent of all species. They are probably derived, with large modifications, from typical chromosomes in the species in which they are found. Bell & Burt (1990) have argued convincingly that they are best thought of as parasites, because they reduce the growth rate and fertility of their 'hosts', but survive in populations because of their ability to increase in number within a lineage. Some examples will illustrate this ability:

- Although B chromosomes usually disjoin at mitosis in the same way as typical A chromosomes, sometimes non-disjunction causes them to increase in numbers in particular cell lineages. In plants, almost invariably they increase in aerial tissues (Jones, 1975). In the smooth hawksbeard, *Crepis capillaris*, an increase in number occurs preferentially in the florets.

- In the grasshopper *Myrmeleotettix maculatus*, B chromosomes move preferentially to the egg pronucleus, rather than to the polar bodies (Hewitt, 1973). This is an important exception to the generalization made above, that genes in meiotic chromosomes are relatively helpless. A variety of other accumulative mechanisms occur in plants (Jones & Rees, 1982).

- A small B chromosome in the wasp *Nasonia vitripennis* is transmitted only in sperm. After fertilization, the paternal A chromosomes (but not the B chromosome) are condensed and ultimately lost, so that the new, haploid, individual becomes a male. In consequence, the B chromosome is transmitted to all the offspring of the new individual.

These examples support the image of B chromosomes as parasites. We would therefore expect the evolution of genes on A chromosomes that suppress these effects. Bell & Burt (1990) reviewed the evidence, still rather sparse, of genetic variation between A chromosomes in their ability to suppress B chromosomes.

10.3 Intrachromosomal repetitive DNA

The control of transposable elements

Orgel & Crick (1980) and Doolittle & Sapienza (1980) suggested that much of the repetitive DNA present in the eukaryotic genome is best regarded as selfish. Since that time, it has become increasingly clear that they were right. But, as the variety and sophistication of these elements has become apparent, a new question has emerged: how is it that higher organisms have not been destroyed by the selfish DNA they carry?

How is the spread of these elements controlled? It is convenient to start with the P factors of *Drosophila* (reviewed by Engels, 1989, and Anxalabehere *et al.*,

1988), because, although they are by no means typical, their regulation is better understood. They were discovered through an analysis of 'hybrid dysgenesis'. When female *D. melanogaster* from long-established laboratory strains were crossed to wild-caught males, there was extensive inviability and infertility among their offspring: no such effects were found in the reciprocal cross. It turned out that this dysgenesis was caused by chromosome elements, of some 3000 bp, present in 30–50 copies in wild flies. The DNA codes for two proteins, a transposase, and a regulatory protein that is present in the eggs laid by females carrying the P factor. Old laboratory strains do not carry the P factor, and hence females from such strains lay eggs that lack the regulatory protein. When chromosomes from wild males are introduced into such eggs, there is extensive transposition of P factors, causing chromosome breakage, and hence inviability and infertility.

Brookfield (1991) has modelled the evolution of this system. He supposes that the population consists of three kinds of factor:

- Autonomous P factors, coding both for a transposase and a repressor protein. These proteins act in *trans*: that is, they influence the transposition of the P factor that coded for them, and of other P elements in the cell.

- Non-autonomous P elements, capable of transposition, and coding for a *trans*-acting repressor, but not for a transposase.

- Non-autonomous P elements, also capable of transposition, but coding for no functional proteins.

Simulations showed that evolution leads to an increase in the frequency of non-autonomous elements: such elements are in fact common in wild populations. Selection also favours the spread of P factors producing a repressor protein. These results depend crucially on the fact that flies in which transposition is frequent are killed or sterilized. Thus, selection between flies, arising from the damaging effect of transposition, outweighs between-element selection favouring transposition. The end result is a population in which transposition is rare, both because many P elements do not code for a transposase, and because of the presence in the egg cytoplasm of a repressor protein coded for by the P factor itself.

P elements are by no means typical of transposable elements. They are not retrotransposons: that is, there is no RNA phase in the replication cycle, as evidenced by the fact that they contain introns. They code for a protein repressing their own replication. Their main effect on fitness is immediate, arising from the process of transposition itself. They are not long-term inhabitants of the *Drosophila* genome: Anxolabehere *et al.* (1988) argued convincingly that they first appeared in *D. melanogaster* in the past hundred years. Thus, old laboratory strains lack them because they never had them, not because they have lost them.

In contrast, copia-like elements are retrotransposons, are not known to code for a repressor, have a less immediate effect on the fitness of their hosts, and are ancient inhabitants of the *Drosophila* genome. Their population biology was reviewed by Charlesworth & Langley (1989). A crucial fact is that there are many potential sites for the insertion of copia (and similarly for other families): most are unoccupied, and it is rare for two unrelated flies to have a copia element at the same site. Transposition rates are low but high enough to cause a steady increase in copy number if there were no counteracting force. What is this force? There are at least three possibilities:

- When an element transposes to a new site, it may cause a mutation, which is likely to be deleterious. Many mutations are so caused: for example, the *apricot* eye-colour mutant was caused by the insertion of a copia element into the *white* gene. But there are reasons to think that mutations do not limit the spread. One is that most copia-like elements are as common on the X chromosome (where they are exposed to selection in males) as on autosomes (Montgomery *et al.*, 1987).

- A repressor protein is produced (coded for either by the fly genome or by the element itself), in amounts increasing with copy number. There is no direct evidence for such a repressor. However, it is worth remembering that the repressor of P-element transposition was discovered because some females lack P elements, resulting in hybrid dygenesis: no flies lack copia.

- If many elements with the same sequence exist in the same genome, this may lead to ectopic recombination (recombination between elements at different chromosome locations) resulting in aneuploid gametes. This is the hypothesis favoured by Charlesworth & Langley. Theoretical analysis (Langley *et al.*, 1988) shows that the process can account for the distribution of elements throughout the genome, in particular the excess of elements in regions of low crossing-over.

It is therefore plausible that these elements are maintained by a balance between transpositional increase, and lowered fertility resulting from ectopic recombination. This cannot, however, be the whole story. The model analysed by Langley *et al.* assumes that all elements have the same transposition rate, and asks what will be the equilibrium frequency and distribution of elements. But suppose that there is variation between elements in rate of transposition. Will not selection lead to an indefinite increase in transposition rate, and hence of copy number?

One possible explanation (Nee & Maynard Smith, 1990) assumes that the molecular biology of retrotransposons resembles that of retroviruses. If so, it is likely that an RNA transcript has two possible fates. One is to act as a template for transposition. The other is to be processed into messenger RNAs for protein synthesis. If sequence differences can influence the fate of a transcript, between-element selection will lead to a high frequency of inactive templates waiting to be

transcribed, and a low frequency of translationally active ones. This would explain why, even though some 3 per cent of polyadenylated RNA in some cell lines consists of copia transcripts, transposition is a rare event. Everyone wants to be a bride, no-one to be a bridesmaid.

There is still much about the evolution of transposons that we do not understand. For example, there is a striking contrast between their distribution in *Drosophila* and in mammals. In *Drosophila*, the frequency of an element at any particular occupiable site is low, and it is rare to find the same site occupied in two unrelated flies. In humans, most elements are fixed, and polymorphism is rare. We do not know why.

If the account just given turns out to be correct, and typical of the regulation of transposable elements, it leads to a rather interesting conclusion. Although selection at the organism level, arising from the harmful effects of transposable elements, is crucial, the actual regulation depends on evolved properties of the elements themselves, and not, as one might have expected, on control by the organism. The reason, of course, is that the continued success of an element depends on the survival and reproduction of the organism. Transposition is favoured mainly because of sexual reproduction. An element that transposes may be inherited from only one parent, but be transmitted to all the offspring. It will be interesting to learn something about the repeated DNA in long-term parthenogens. The importance of the mode of transmission between organisms will appear again in our discussion of symbiosis in the next chapter.

Repetitive DNA and the C-value paradox

It is bad for human self-esteem that it should, apparently, require 40 times as much DNA to make a lungfish as it does to make a man or woman. This is an example of the 'C-value paradox' (Thomas, 1971)—the C-value being a term for the DNA content of a cell. Now that we know that the reason for the discrepancy is the greater amount of non-coding DNA in lungfish, the paradox no longer seems so threatening, but the question remains: why is the amount of non-coding DNA so great, and so variable?

Since there is little non-coding DNA in prokaryotes, we confine our discussion to eukaryotes. Briefly, the facts are as follows (Cavalier-Smith, 1985). The greatest range of variation occurs in unicellular eukaryotes: the range from yeast to amoeba is 80 000-fold, and within green algae is 3000-fold. In contrast, the range of variation in classes of multicellular animals and plants is 10- to 100-fold. The three classes of amniote vertebrates—reptiles, birds and mammals—are the most uniform (2- to 4-fold). There is a striking correlation between DNA content, nuclear volume and cell volume, the relations being isometric. The cell-volume correlation holds only for growing, dividing cells: it does not hold for post-mitotic cells, or for embryonic cells in periods of development when there is cell division but little or no protein synthesis. Also, the correlation is not perfect.

Haploid and diploid hymenopterans have cells of the same size (Darlington, 1939), although sporadically occurring haploids in diploid species have cells of roughly half the typical size, suggesting that in hymenopterans gene mutations have been selected that alter the DNA/cell volume ratio. The possibility of such mutations is confirmed by the fact that, in both fission and budding yeasts, single mutants are known that reduce cell volume, approximately by one half, by altering the size at which the cell divides (Nurse, 1985).

How are these facts to be explained? Is there a direct physiological connection between DNA content, nuclear volume and cell volume, or are the correlations maintained by selection? If the latter is the case, is selection acting on the organism, or on selfish DNA elements, or both? It is convenient first to outline the explanation offered by Cavalier-Smith (1985 and earlier), and then to consider an alternative hypothesis. In brief, he suggests:

- There is a direct physiological connection between DNA content and nuclear volume, perhaps arising from the structural properties of condensed chromosomes, and their attachment to the nuclear envelope by their centromeres and telomeres. On this hypothesis, variations in nuclear volume for a given DNA content arise from genically controlled variation in the degree of chromosome condensation.

- There is an approximately constant relationship, in growing cells, between nuclear volume and cell volume, which is maintained by selection, because there is an optimal ratio between the space allocated to transcription and RNA processing in the nucleus, and to translation in the cytoplasm. This ratio depends on properties common to all eukaryotes, and should therefore be constant.

- Selection at the level of the organism alters cell volume: this in turn selects for altered nuclear volume, which is brought about by changes in the amount of non-coding DNA. There are various ways in which cell size can affect organism fitness. Small cell size is observed to correlate with high growth rate. In animals, large cells are associated with a low metabolic rate. The lungfish *Protopterus* and *Lepidosiren* aestivate, and have the largest genomes of any animal. The fossil evidence shows that osteocyte size has increased in lungfish during the past 400 million years (Thomson, 1972). In plants, Grime (1983) found that plants with the most rapid leaf elongation in spring have the largest genomes: presumably, rapid growth at that time does not depend on cell division. No doubt there are many other such selective effects.

To summarize Cavalier-Smith's argument, then, selection on organisms alters cell volume. Changes in cell volume impose selection for altered DNA content, in order to maintain an appropriate ratio between cell and nuclear volume. He has, therefore, no need to postulate intragenomic conflict. If, however, there is selection at the genomic level leading to increased DNA content per cell, this could lead to a nuclear volume greater than that optimal for a given cell volume, or to

cells larger than the optimum. It is plausible that there should be an increase in genome size, driven by selection for more efficient transposition. In a sexual population, repeated elements that are particularly efficient at replication and transposition will increase in number. We now describe a case in which DNA content and cell volume may have been driven well above the level that would be optimal for the organism.

The evolution of chromosome form

It is a curious fact about evolutionary biology that, despite the vast body of data and many millions of words, little sense has been made of the evolution of chromosome form. We suspect that this is because we have been looking for sense in the wrong place. We start, therefore, by describing in some detail a particular case, the plethodontid salamanders, in which something is known of both morphological and karyotypic evolution. In this case, it seems that evolution has been driven by selection at the gene level, but that selection at the organism level has brought about compensating changes. This conclusion may be of more general application.

In these salamanders there is, as expected, a correlation between large genome size, large cell size, low metabolic rate and slow development (Roth *et al.*, 1988). In the tribe Bolitoglossini, a tropical group of plethodontids, this is associated with an extreme reduction in body size. *Thorius pennatulus* may be the smallest tetrapod vertebrate, with a snout to vent length of 15 mm, and extremely slender. Paradoxically, the bolitoglossines have genome sizes that are among the largest among animals (25 pg of DNA per haploid genome in *T. pennatulus*, and up to 75 pg in some members of the tribe). They have a highly specialized feeding mechanism, with a projectile tongue and binocular vision that provides accurate depth perception, enabling them to feed on minute mites and collembola (<0.5 mm in body length). To maintain this feeding mechanism, despite simultaneous miniaturization and increase in cell size, has required extensive anatomical changes. The brain and eyes have increased relatively in size (12% and 10% of head volume, respectively, in *Thorius*, compared to 4% and 7% in a larger species in the same family). Brain anatomy is simplified. The tectum, concerned with the analysis of visual input, is larger relative to other parts, especially the forebrain, and the proportion of grey matter (cell bodies) to white (nerve fibres) is increased.

The large genome size of the bolitoglossine salamanders is almost certainly derived, and not primitive, although there has probably been a secondary reduction in *Thorius*. There is a similar association between large genome size and simplified brain anatomy in frogs. Given the increase in genome size, and consequently of cell size, the anatomical changes make sense. But why the increase in genome size? It is hard to resist the conclusion that the increase has been driven by selection, at the gene level, for successful transposition.

Macgregor (1982) summarized what is known about the cytology and molecular biology of another lungless salamander, *Plethodon*. The genus, which has existed for some 80 million years, now has 45 known species, all in North America. These are extremely similar in morphology, differing mainly in size, geographical distribution and details of colour. They do vary, however, in C-value, from 20 to 66 pg of DNA. The karyotypes are otherwise identical: it is as if all the chromosomes had expanded uniformly. Most of the genome consists of middle repetitive DNA. Macgregor estimated that there are some 6000 copies each of some 250 sequences, each 100–200 nucleotides long. The sequences are 90 per cent similar in closely related species, but only 10 per cent similar in more distantly related species. This is just the pattern one would expect if, from time to time, a new, highly transposable, sequence arises, and spreads randomly throughout the genome: there must also be some process that stops an indefinite increase in copy number.

Macgregor himself thinks that there is a direct causal link between constancy of karyotype and of morphology (although he has evidence that there have been changes in the location of particular genes—in particular, of ribosomal genes). This seems to us unlikely. Certainly, karyotypic change does not necessarily entail morphological change: for example, two phenotypically similar species of deer, *Muntiacus reevesii* and *M.muntjac*, have $2n = 46$ and $2n = 6$ or 8, respectively. The situation in *Plethodon* seems most likely to be the result of the particular types of repetitive DNA present. Although we do not know that the molecular biology is the same for the plethodontids in general, the work of Roth and his colleagues, described above, suggests that the changes in genome size occurred autonomously, and that selection at the organism level has produced the anatomical changes needed to compensate for the changes in cell size.

A similar explanation may prove to be true for karyotypic evolution in general. It is perhaps natural that scientists who spend their lives looking at the shapes of chromosomes should come to think that these shapes have some significance for the organism, but there is little evidence for this view. It is more likely that the pattern of karyotypic evolution depends mainly on the types of repetitive DNA present, which in turn depends on selection at the genic rather that the organismic level. A second relevant fact is that some types of karyotypic change may be tolerated by some types of organism more readily than others. For example, inversions are very frequent in *Drosophila* and other Diptera, because they do not cause infertility, as they do in other organisms: there is no recombination that might give rise to aneuploid gametes in males, and, in females, aneuploid nuclei are eliminated in the polar bodies.

10.4 Avoiding conflict between organelles

Grun (1976) suggested that uniparental inheritance evolved because it prevents the spread of selfish organelles. Since the division of organelles—mitochondria

and chloroplasts—is usually not tightly synchronized with the division of the nucleus, there is a clear danger of conflict. A mutant organelle that replicated more rapidly, at the expense of performing its cellular functions less efficiently, would increase in frequency, and in so doing would lower the fitness of nuclear genes. The likelihood of such a conflict is greatly reduced in practice because organelles are usually inherited only from one parent (see Whatley, 1982, for a review of plastid inheritance in plants). Hence, organelles in a cell are genetically similar and selfish mutants are eliminated before they can spread through the population.

The interesting question is whether uniparental inheritance of organelles is merely an unselected consequence of the unequal cytoplasmic contribution of the two sexes, which itself evolved for other reasons, or whether it was selected for in its own right. Some reasons for thinking that selection has been involved are as follows:

- Uniparental inheritance occurs also in isogamous organisms. In the green alga, *Chlamydomonas*, the chloroplast is inherited only from the parent of $+$ mating type (mt^+), and the mitochondria from the $-$ mating type (mt^-). The genetic mechanism (Sager, 1977) is of some interest, particularly because of the involvement of nuclear genes, including a gene in the mt^+ nucleus coding for an enzyme that degrades its own mitochondrial DNA. In *Chlamydomonas*, the destruction of chloroplast DNA from one parent occurs only after fusion of the two chloroplasts. This is unusual: a commoner pattern is for the plastids from one parent to disintegrate.

- In some cases, cytoplasm is shed from the male gamete before fertilization. For example, in the fern *Mausilea vestita* (Myles, 1975) most of the male cytoplasm is shed before fusion: first, an anterior cap containing most of the mitochondria, and later a posterior vesicle containing the plastids, are pinched off. This does not look like an unselected consequence of unequal gamete size.

- In conifers, plastids are usually inherited from the male parent. (In passing, this may be the main reason why conifers seem never to give rise to parthenogenetic variaties.)

10.5 Distortion of sex allocation

Fisher (1930) argued that, if the sex ratio is determined by autosomal genes in a parent, selection will lead to an equal allocation of resources to the two sexes: typically to a 1:1 sex ratio. Intragenomic conflict, however, can lead to distortion (Hurst, 1993, gives an excellent review).

We saw in section 10.2 that meiotic-drive genes on the sex chromosomes can distort the sex ratio. Once the sex ratio is distorted, however, selection will

favour autosomal genes that suppress distortion. The sex ratio can also be distorted by genes on organelles that are uniparentally inherited: the examples discussed below are taken from Hurst's review.

A classic example of intragenomic conflict is afforded by cytoplasmic male sterility in flowering plants (Lewis, 1941). The phenomenon is common. Most flowering plants are hermaphrodites. Theory predicts that hermaphrodites should allocate resources equally to male and female function (Maynard Smith, 1971). Although this is hard to test quantitatively, there is plenty of evidence, going back to Darwin (1877), that a hermaphrodite that ceases to allocate resources to pollen can produce more seeds. Therefore, a gene in an organelle that is transmitted only in the ovule will be favoured by selection if it suppresses male function. In fact, many plant species consist of a mixture of hermaphrodites and male-sterile individuals: the condition is known as gynodioecy. In *Thymus vulgaris*, both genetic and population studies have been carried out (Couvet *et al.*, 1986). Male sterility is caused by alleles in the mitochondria, and is suppressed by nuclear genes. Rather surprisingly, male sterility has not been reported in hermaphroditic animals.

In dioecious species, genes in maternally transmitted organelles, or parasites, can increase their fitness in several ways:

By converting sexual females into parthenogens. A cytoplasmic gene transmitted to a son has no future. Stouthamer *et al.* (1990) reported that asexual strains of a solitary wasp, *Trichogramma*, reverted to bisexual haplodiploids when treated with bacteriocidal antibiotics, suggesting that asexuality is caused by an endosymbiont.

By converting males into females. For example, Bulnheim (1978) reported that female amphipod crustaceans, *Gammarus duebeni*, infected with either of two protozoan symbionts, *Octospora effeminans* and *Thelohania hereditaria*, produce all-female broods, the condition being inherited by the daughters. Experimental infection by *Octospora* is effective in inducing all-female broods.

By killing males early in their lives. If males compete with their sisters, an endosymbiont that kills a male in which it finds itself will increase the survival chances of its relatives in sisters of that male. Several examples of this phenomenon have been reported. The best understood is the 'sex ratio' (SR) condition in neotropical *Drosophila* of the *willistoni* group (Malagalowkin & Poulson, 1957). The causative agent was first thought to be a spirochaete, but is now ascribed to a rather enigmatic group of prokaryotes, the spiroplasms (Whitcomb, 1981). Infected females produce all-female broods, because male eggs die. The condition is inherited by daughters. The agent has been grown in cell-free culture, and is still effective when re-injected. Hurst (1993) reviewed a number of cases of male-killing by symbionts. He distinguished between early killing agents, such as SR, and those that kill males later in life, when the males no longer compete with their sisters: in the

latter cases, he argued that the advantage to the parasite is that, after killing its male host, it can be transmitted horizontally via a secondary host.

Of the examples discussed, only cytoplasmic male sterility in plants is strictly one of intragenomic conflict: the others involve conflict between a host and its parasites. The distinction is that *Thymus* could not survive without its mitochondria, whereas *Trichogramma*, *Gammarus* and *Drosophila* can live quite happily without their parasites. However, the parasite examples have in common that the agents are uniparentally inherited, as are many organelles. Perhaps what a gene in a parasite can do, a gene in an organelle can do equally well.

11 Symbiosis

11.1 Introduction

The establishment of a permanent and obligate coexistence of genetic entities that were once capable of independent existence played an important part in the origin of the eukaryotes, and, if our earlier speculations are correct, in the origin of cells and chromosomes. In this chapter, we discuss other examples of symbiosis. The term is used to include all cases in which two or more different kinds of organism live in close association: thus it extends from parasitism to mutualism. Mutualism has been defined as a relationship from which both partners benefit. However, as will become clearer below, it is hard to measure, or even to define, 'benefits': in what sense is a mitochondrion today better off than its once free-living ancestors? The two questions that we shall ask are:

- What are the selective force acting on the two partners in present-day symbioses?
- Could such selective forces lead to the establishment of permanent and obligate coexistence?

First, however, we review briefly some of the ecologically more important symbioses (for further examples, see Pirozynski & Hawksworth, 1988; Margulis & Fester, 1991). We mention only a fraction of the known mutualistic associations. Others, including cases of interaction between animals and prokaryotes, are discussed below. It is striking that symbiotic relationships have been important in the utilization by plants of nutrient-poor soils, the colonization of bare rock, life in deep-sea vents, the construction of coral reefs, and the utilization of plant material by several groups of insects.

11.2 The ecology of symbiosis

Prokaryotes

Sonea (1991; see also Sonea & Panisset, 1983) has pictured the world of bacteria as a single superorganism, whose individual component cells rely for their survival on ecological exchange of metabolites, and on genetic exchange via plasmids and phages. This picture has the virtue of emphasizing the important role played by plasmids and temperate phages in conferring on individual bacterial cells capacities needed in particular environments—for example, resistance to antibiotics, tolerance of heavy metals and new metabolic abilities. But the picture suffers from the drawback that is fatal to all holistic models of evolution, from the Gaia hypothesis downwards, of losing all sight of the units of selection, and hence of lacking any model of the dynamics of evolutionary change. A bacterial genome can indeed be seen as a temporary alliance between the bacterial chromosome, and one or more small accessory replicons that provide

additional functions. However, the bacterial chromosome itself is of extra-ordinary stability. This is demonstrated by the fact that large-scale phylogenies based on different chromosomal genes are very similar (Woese, 1987), which would not be the case if horizontal gene transfer was at all common. To give a more specific example: *Salmonella* and *Escherichia* are thought to have evolved independently for many millions of years—perhaps since the divergence of their main hosts, reptiles and mammals. Yet their chromosomal linkage maps are almost identical, and sequence analysis of over 100 genes shows that gene exchange, if it has occurred at all, has been rare (Sharp, 1991; Maynard Smith, unpublished data).

It is therefore better to see the bacterial world, not as a superorganism, but as a number of independently evolving chromosomal lineages, exchanging genes and parts of genes with close relatives, and forming temporary symbiotic re-lationships with a vast array of small and often very unstable replicons. These temporary symbioses are probably the main process whereby bacteria adapt rapidly to local circumstances.

Symbioses between animals and algae

Many aquatic invertebrates have endosymbiotic green algae, from which they acquire nutrients. The algae are usually capable of independent existence outside their hosts, and must be acquired afresh by each individual host. A few kinds of algae are found in taxonomically very diverse hosts: for example, only two genera of dinoflagellates are known to live within marine invertebrates be-longing to three phyla—Cnidaria, Mollusca and Platyhelminthes. The associa-tions can be ecologically important: for example, coelenterates can build coral reefs only with the aid of dinoflagellate algae.

Symbioses between plants and fungi

Lichens are fungi with endosymbiotic 'algae', which may be cyanobacteria or green algae (for a review, see Hawksworth, 1988). Some 13 000 known species of fungi, belonging to over 500 genera, live in lichen associations. Most are ascomycetes, among which the symbiotic habit has evolved many times. The algae are much less varied, belonging to fewer than 50 genera. All the cyano-bacteria involved in such associations also occur free in nature, as do most of the green algae, although two of the commonest genera, *Myrmecia* and *Trebouxia*, are rarely found free-living. Sometimes it seems that the fungus is merely ex-ploiting the algae: for example, the fungus *Collema* can invade a colony of free-living algae (*Nostoc*) and mould it into a characteristic lichen thallus. In other cases, the association is more permanent, and joint propagules, usually con-taining many nuclei of both partners, are produced.

Of even greater ecological significance is the association between mycorrhizal fungi and land plants (for reviews, see Lewis, 1991; Kendrick, 1991). The most primitive group of mycorrhizal fungi, the vesicular-arbuscular (VA) fungi, are

known as fossils from the Devonian, and may be older. They are obligate symbionts, whose hyphae penetrate the cells of the plant root (i.e. they are 'endomycorrhizae'). They are asexual: since much of fungal taxonomy depends on sexual structures, their taxonomic position is uncertain. Their role in the initial colonization of the land would have been to enable plants to exploit essentially mineral soils. Today, they are the dominant mycorrhizae in such soils, particularly in the tropics. There is a net flux of minerals from fungus to plant, and of carbon from plant to fungus, so the association can reasonably be regarded as mutualistic. The more recently evolved ectomycorrhizal fungi, which do not penetrate plant cells, are mainly Basidiomycetes. They are associated with plants growing in more organic soils at high latitudes and altitudes. Of particular interest are the ericoid fungi, associated with ericaceous plants growing in acid soils. In addition to providing minerals, they make available nitrogen from organic compounds that would otherwise be unavailable to the plant. There are very few species of ericoid fungi—perhaps only one—associated with three families of plants. They can be found free-living, and are readily cultured, in contrast to the more ancient VA fungi, which cannot grow except in association with plants.

Associations between animals and fungi

One of the most dramatic examples of mutualism is the farming of fungi by insects. In central and south America, leaf-cutting ants (*Atta*) cut a harvest of leaves, which they masticate underground, where they are digested by various species of cellulase-possessing fungi. The fungal mycelium then constitutes the food for the ant colony. Not only have the ants evolved complex behaviours ensuring the survival of the fungi: the fungi have evolved inflated tips to their hyphae (gongylidia), which serve as food for the ants. Most termites can use cellulose with the help of a varied population of microorganisms in the hind gut, but the mound-building termites of Africa and Asia, of the family Macrotermitinae, lack such microorganisms, and resemble *Atta* in culturing fungi.

11.3 A model

Most cases of symbiosis involve interaction between a host organism and a symbiont living inside or on the surface of the host. The symbiont can range from a damaging parasite to a necessary mutualist. To understand their coevolution, we must consider variation in both partners. A simple model, based on one analysed by Law (1991), is shown in Fig. 11.1. Numerical values of the fitnesses have been used, because we are interested only in inequalities, not absolute values. Note that only the relative fitnesses of different types of host, or of symbiont, are relevant: comparisons of the fitness of host and parasite are irrelevant. For simplicity, we consider only two discrete strategies for the host, and two for the symbiont, although variation could in practice be continuous.

The assumptions made in Fig. 11.1a are as follows:

1. If symbionts cooperate, it pays the host to cultivate them.

2. If symbionts are parasites, it pays the host to kill them.

3. The host always does better if the symbionts cooperate than if they are parasites.

4a. Whatever the strategy of the host, it pays the symbiont to parasitize.

The first three of these assumptions can be regarded as definitions of the terms 'cooperate' and 'parasitize'. Assumption (4), however, will sometimes be false. In particular, if the host is a cultivator, it might pay the symbiont to keep the host alive: it does not pay to kill the goose that lays the golden eggs. The assumption is therefore modified in Fig. 11.1b and 1c, as follows:

4b. If the host is a cultivator, it pays the symbiont to cooperate: if the host is a killer, it pays the symbiont to parasitize.

4c. Whatever the host strategy, it pays the symbiont to cooperate.

The evolutionary outcome of Fig. 11.1a is parasitism, with the host evolving defense mechanisms. It is reminiscent of the Prisoner's Dilemma game (see section 16.2), in that the evolutionary outcome is unfavourable for both participants. For Fig. 11.1b, there are two stable states, parasitism and mutualism. For Fig. 11.1c, the outcome is mutualism.

In deciding which of these diagrams is appropriate in particular cases, the crucial feature is the way in which symbionts are transmitted between generations (Fig. 11.2). We list some of the possibilities.

Each individual host acquires some genetically different symbionts from the environment (Fig. 11.2a). In this case, it is likely that Fig. 11.1a is appropriate. There is no advantage to an individual symbiont in keeping the host alive by cooperation if it is going to be killed by other, parasitic symbionts. More generally, if there is genetic variation among the symbionts, it is likely that there will be some phenotypic change that would enable a 'selfish' symbiont to invade a population of cooperators. Initially, this might be no more than a phenotype corresponding to what economists call a 'free-rider': that is, a symbiont that does not waste energy performing useful, cooperative actions. Satellite viruses are such free-riders in the case of parasitism: their presence is likely to benefit the host, or rather, to reduce the damage to the host. If symbionts are cooperative rather than parasitic, the presence of free-riders would reduce the benefit to the host: Nealson (1991) gives an example of such free-riders in a complex case of symbiosis. Often, of course, a 'selfish' symbiont will cause active damage to the host.

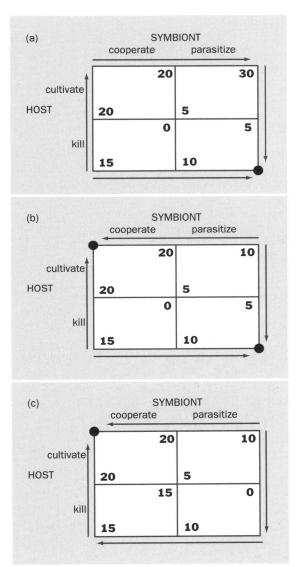

Fig. 11.1 A model of the evolution of symbiosis. The entries in each square of the tables represent the fitness of (1) bottom left, host individuals adopting the strategy on the left, if the symbiont adopts the strategy above; and (2) top right, symbiont individuals adopting the strategy above, if the host adopts the strategy on the left. •, Stable states of the system. (Modified from Law, 1991.)

Each host acquires a single symbiont (or clone of symbionts) from the environment (Fig. 11.2b). The best strategy for the symbiont is now hard to predict. If the host is a cultivator, it is likely to pay the symbiont to cooperate, on the 'golden egg' principle. If the host adopts the killer stratgy, it may pay the symbiont to multiply as rapidly as possible, regardless of any damage to the host, before the killing response is effective, giving the fitnesses shown in Fig. 11.1b.

Symbiosis

Fig. 11.2 Modes of transmission of endosymbionts. (a) Indirect transmission, with several symbiont genotypes infecting each host; (b) indirect transmission, with only one symbiont genotype infecting each host; (c) direct transmission.

But if the symbiont can resist the killing strategy of the host, it may still pay it to minimize the damage to the host, as in Fig. 11.1c.

That is the case analysed by May & Anderson's (1983) model of the evolution of myxomatosis. The symbiont will evolve the best compromise between excessive virulence, which kills the host and the parasites along with it, and insufficient virulence, which reduces the rate of infection of new hosts. It is hard to make general predictions, because the outcome will depend on the detailed biology. However, with hosts infected by a single symbiont, cooperative mutu-

alism is likely to be stable once it evolves. The evolution from parasitism to mutualism will be favoured if the host's killing response is ineffective, and if the further spread of the symbiont is greater if the host does survive. It will not occur if the host can rapidly rid itself of the parasite, or if the parasite spreads only by killing the host.

Each host acquires one or a few symbionts from one of its parents (Fig. 11.2c). Given such direct transmission, the symbionts in a host will be genetically identical, except for recent mutation. If every host egg (or other propagule) contains symbionts, we expect mutualism to evolve. However, since the transmission of host and parasite genes are not identical, possibilities of conflict of the kinds discussed in Chapter 10 still arise.

There are other possibilities: for example, transmission of symbionts from both host parents, or transmission of a large number of symbionts from one host parent. However, when reviewing the data on modes of transmission, the following theoretical conclusions are worth bearing in mind:

- With multiple transmission of symbionts from the environment, parasitism is the expected outcome. Cooperation will evolve only if there is no 'selfish' phenotype available to the symbiont.

- With single transmission of symbionts from the environment, evolution can lead to mutualism or parasitism. Mutualism is favoured if the killing response of the host is relatively ineffective, and if symbiont transmission is favoured by long-continued host survival.

- With direct transmission of one or a few symbionts, by one sex of the host, mutualism is the expected outcome.

11.4 Modes of transmission

The grain weevil, *Sitophilus oryzae*, has endosymbiotic Gram-negative bacteria (Nardon & Grenier, 1991). Although the weevil can develop and reproduce without the bacteria, it does so much less well. The bacteria have not been cultured outside the beetle. They are transmitted in the female germ line only. They supply five vitamins, including biotin and pantothenic acid. The intracellular bacterial symbionts that are essential for the survival of aphids provide an even clearer case of direct transmission in the female line. Munson *et al.* (1991) sequenced 16S RNA from the bacteria and found that the symbionts from four families of aphids have a phylogeny that is parallel to that of the aphids themselves.

Although such direct transmission does occur, it seems to be the exception rather than the rule. As mentioned above, some lichens produce multicellular propagules, but in others the fungus must re-acquire algae afresh in each generation. Similarly, plants must re-establish root connections with mycorrhizae in

each generation. Some animals are fully dependent on symbiotic bacteria and must swallow them as larvae. For example, luminous bacteria live in the light organs of many teleost fish. In the best-studied group, the leiognathids, the bacteria seem to be re-acquired in each generation, because the eggs do not carry the bacterium, which can be cultured from the seawater in which the fish are found (McFall-Ngai, 1991). The vestimentiferan worm *Riftia pachyptila*, an inhabitant of deep-sea vents, has no mouth or anus as an adult. It obtains its energy from symbiotic bacteria, confined to special organs, that oxidize sulphides present in the vents (Vetter, 1991). Sulphide and oxygen are transported separately to these organs in the blood, bound to a special haemoglobin (the sulphide is bound to the globin, and the oxygen to the haem). The larval *Riftia* has a mouth, gut and anus, and acquires the bacteria by feeding.

An intermediate pattern of transmission holds for the gut microorganisms of termites. The grub acquires them by feeding at the anus of an adult. Because of the closed nature of the colony, this means that a new individual acquires its symbionts from a relative, and not randomly from the environment, as in luminous fish and in *Riftia*.

11.5 Irreversibility

Although, as we have seen, mitochondria do have selfish options—for example, in distorting the sex ratio—they do not have the option of reverting to an independent free-living existence. Some other endosymbionts have reached a similar stage of irreversibility. Vesicular-arbuscular mycorrhizae have not been cultured apart from plants: in this, they contrast with more recently evolved mycorrhizae. Similarly, the bacteria symbiotic in aphids have not been cultured. A more extreme stage has been reached by *Cyanophora paradoxa*, a euglenoid-like protist that has lost its chloroplast, and has subsequently acquired an endosymbiotic cyanobacterium (Trench, 1991), which has a genome of only 128 kb, and which imports 80 per cent of its constituent proteins from its host. Thus, it is not only the original organelles—mitochondria, chloroplasts, and perhaps others—that have lost for ever the ability to survive outside their hosts.

11.6 Does symbiosis evolve towards mutualism?

Until recently, the conventional view on the coevolution of hosts and symbionts was that expressed by Dubos (1965): 'Given enough time a state of peaceful coexistence eventually becomes established between any host and parasite.' This view was challenged by May & Anderson (1982), who pointed out that the outcome of coevolution depends on the precise relationship between virulence and infectivity. Further, if the conventional view is correct, it is hard to understand why there are so many virulent parasites around, or why hosts have

gone to so much trouble to evolve resistance to them. May & Anderson supported their conclusion by an analysis of the data on myxomatosis in rabbits, quoted above. In a parallel critique of Dubos' position, Ewald (1983) compared the severity of disease in humans caused by parasites typically infecting humans with that of closely related parasites usually found in animals, but occasionally infecting humans. For example, the human malarial parasite, *Plasmodium vivax*, causes more severe symptoms in man than does the closely related *P. cynomolgi*, a parasite of macaques that is occasionally transmitted to humans. There are, of course, examples in which the difference is in the other direction, but Ewald claimed that examples resembling the one just given are commoner, and certainly common enough to disprove the view that coevolution necessarily leads to benevolence in parasites.

The importance of horizontal as opposed to vertical transmission was stressed by Maynard Smith (1991), and by Bull *et al.* (1991). The latter authors supported their theoretical argument by an experiment on the coevolution of the bacterium, *E. coli*, and a filamentous DNA phage. They showed that, if conditions were such as to permit only vertical transmission, the successful phages were those that had the least deleterious effect on the host's growth rate: with horizontal transmission, phages with more harmful effects were selected.

The prediction that horizontal transmission should be associated with greater virulence is clear. Comparative data seem to confirm it. It is interesting to compare the biology of the plasmids and phages symbiotic in bacteria. Some plasmids have vertical transmission only; others can also be transmitted horizontally by conjugation. When horizontal transmission occurs, the donor bacterium is not damaged. In contrast, virulent phages can only be transmitted horizontally: temperate phages also have a phase in which their DNA is incorporated into the host chromosome, and transmission is vertical. Horizontal transmission always involves the destruction of the host bacterium.

These different patterns of transmission are associated with different degrees of mutualism or parasitism, in the expected way. Plasmids often carry genes that benefit their hosts, by conferring resistance to antibiotics, drugs or heavy metals, by producing toxins, by coding for restriction enzymes or by enabling the bacterium to utilize novel substrates. Occasionally, however, plasmids may destroy their hosts, as in the case of the ColE1 plasmid. Temperate phages also may confer some benefit on their hosts when integrated into the host chromosome, but all phages destroy their hosts prior to horizontal transmission. Not surprisingly, bacteria are able rapidly to evolve resistance to phage infection, usually by altering the cell surface so that phage particles cannot infect.

Turning to eukaryotic hosts and parasites, confirmation is provided by Herre's (1993) study of the nematode parasites of figwasps. Mature female wasps enter the fig inflorescence, pollinate the flowers, lay eggs and die. The dispersive phase of the nematodes enter the body cavity of newly emerged females, and are carried by them to new figs. The nematodes consume the body of the wasp,

emerge, mate and lay eggs in the same fig. Hence, if only a single wasp foundress enters a fig, any nematode she carries will produce offspring that infect her offspring: that is, transmission is vertical. But if several female wasps enter a fig, a nematode in one wasp female may produce offspring that infect the offspring of a different female: transmission is horizontal as well as vertical.

Particular fig species have their own pollinating wasp species, and those in turn have their own nematode species. Herre studied 11 wasp–nematode pairs. The proportion of single-foundress broods varied from 1 to 0.2. In species with a single foundress, nematode infection caused no reduction in a female's life-time reproductive success, whereas in species with multiple foundresses, the reproductive success of infected females was about 80 per cent of that of uninfected ones. It is interesting that the single-foundress species also had a more female-biased sex ratio (as predicted by Hamilton, 1967), and less-aggressive males. On the basis of phylogenetic studies, Herre argues that the association is ancient, and started as a commensal one: subsequent evolution has been towards parasitism, at least in those species with horizontal transmission.

Evolution in the opposite direction is suggested by the work of Clay (1988) on the ergot-producing fungi infecting grasses. Fungi of the genera *Balansia* and *Epichloe* are systemic endophytes of grasses. They are sexual, and produce dispersive spores. The ergot that they produce protects the host plant against mammalian and insect herbivores, but the fungi sterilize the host plant, and so must be regarded as parasites. However, some fungi in the related genus *Acremonium*, also endophytic and ergot-producing, do not sterilize their hosts. Interestingly, they are asexual, and transmitted in the host seed. Again, there is an association between vertical transmission and lack of virulence. However, in this case it seems clear that evolution has been from horizontal transmission and parasitism towards vertical transmission and mutualism.

Thus there are data to support the view that the mode of transmission, vertical or horizontal, will influence evolution towards mutualism or parasitism, respectively. But plenty of puzzles remain. For example, the transmission of luminous bacteria in fish and of sulphur bacteria in vestimentiferan worms is horizontal, yet the symbionts are essential to the survival of their hosts.

11.7 Evolution within a host

Our discussion of the coevolution of hosts and symbionts has rested on two assumptions:

- If there are genetic differences between the symbionts within a host, these arose because the host was infected by more than one symbiont: genetic differences between the descendents of a single infecting symbiont can safely be ignored.

- There may be a positive correlation between high virulence and effective transfer to a new host. Selection for successful transmission is therefore likely to lead to increased virulence.

As we are learning to our cost in the case of HIV, however, the first assumption is not always true: extensive evolution can occur within a single host, from an initially uniform infection. This is an extreme case, because the high mutation rate of RNA viruses leads to very rapid evolution. Within-host evolution, however, is likely to be a general phenomenon. B. R. Levin & J. J. Bull (in an unpublished manuscript) have pointed out that it can lead to an otherwise puzzling association between high virulence and low transmission. Consider, for example, the bacterium *Neisseria meningitidis*, the cause of an often fatal meningitis. The typical habitat of this bacterium is the human throat, where it causes no adverse symptoms, and is readily transmitted to new hosts. To cause meningitis, it must first enter the bloodstream, and then cross the blood–brain barrier. In the short term, this is an excellent habitat for the bacteria, but, no matter how rapidly they multiply, bacteria in the cerebrospinal fluid will never infect a new host. There is an association between high virulence and zero transmission. The explanation is that natural selection within the host produces changes that are advantageous in the short run, but fatal, to the bacterium, in the long run.

11.8 Symbiosis, variability and sex

Law & Lewis (1983) make two interesting points about mutualists. First, the taxonomic diversity of mutualists is small compared to that of their hosts. Some examples illustrating this point were mentioned in section 11.1: only two genera of dinoflagellates live in three phyla of marine invertebrates; relatively few algal species form lichen associations with many species of fungi; and only one (or perhaps several) species of ericoid mycorrhizal fungus is associated with three families of flowering plants. Their second point is that mutualists are often asexual. Again, examples have already been mentioned. The vescicular-arbuscular mycorrhizal fungi are asexual, as are the much more recently evolved ergot-producing mutualists in the genus *Acremonium*. It is easy to suggest why this should be so. Different host species may converge, evolutionarily, on a single mutualist symbiont, whereas they will disperse in phenotypic space to escape the attacks of a parasite, thus encouraging speciation in the latter. Parasites must retain sex so as to evade the resistance mechanisms of their hosts, whereas mutualists can afford to maintain a constant phenotype to which their hosts will accommodate. But these are hand-waving explanations: a more formal model is needed. Also, there are plenty of species-specific, sexual mutualists.

12 Development in Simple Organisms

12.1　The origins of development

Complex multicellular organisms, whose bodies consist of differentiated cells of many kinds, have evolved independently on three occasions—animals, higher plants and fungi. In addition, multicellular organisms with a lesser degree of cellular differentiation have evolved on a number of occasions. For example, the algae have given rise to 'seaweeds' several times. In this and the next three chapters, we discuss the origin and subsequent evolution of such organisms.

Some 540 million years ago, at the beginning of the Cambrian, there appeared an array of multicellular marine animals, including the major phyla that exist today—coelenterates, platyhelminths, annelids, arthropods, molluscs, echinoderms and others. Chordates are also present in the Cambrian: they are not known from the earliest deposits, in which only hard parts are preserved, but are present in the slightly later Burgess Shale, in which soft-bodied forms are preserved. Forty years ago, this sudden appearance of metazoan fossils was not only a puzzle but something of an embarassment: the absence of any known fossils from earlier rocks was a weapon widely used by creationists. Today, the fossil evidence for prokaryotes goes back 3000 million years, and for protists some 1000 million years.

The Cambrian explosion remains a puzzle, however, which has been only fitfully illuminated by the discovery of the enigmatic soft-bodied Ediacaran fauna, which had a worldwide distribution between 580 and 560 million years ago. There are still doubts about how these fossils should be interpreted (Conway Morris, 1993). The orthodox, and more plausible, view is that the fauna is dominated by coelenterates, with some specimens identified as echinoderms and annelids. An alternative interpretation (Seilacher, 1992) is that they belong to an extinct clade of multicellular eukaryotes, the ventobionts, probably lacking an alimentary canal, muscles and nervous system. Although such organisms may have existed, at least some of the Ediacaran fauna have been successfully compared to recent metazoans. If the interpretation of most of these fossils as coelenterates proves to be correct, it would fit in well with the morphological and molecular evidence. The molecular data suggest that coelenterates arose early, but probably not independently of other metazoans. Morphologically they are simple in being diploblastic (formed from two cell layers), in contrast to the triploblastic animals that predominate in the Cambrian.

The puzzle is why the Cambrian explosion took place when it did. Two kinds of answer are possible. One is that, before complex multicellular organisms could evolve, some crucial invention or inventions in cell physiology or gene regulation had to be made: once made, there was rapid radiation into an ecologically empty world. The apparently monophyletic origin of the Metazoa, deduced from molecular data, is consistent with this view. In this and the next three chapters, we discuss what such inventions might be. An alternative is that it was not until some 600 million years ago that physical conditions on the Earth made the

metazoan way of life practicable. A plausible explanation for the origin of hard parts—such as shells and carapaces—at the beginning of the Cambrian is that the first effective predators evolved at that time. But that is an explanation of hard parts, not of multicellular bodies. The most likely explanation for the latter is a rise in the concentration of atmospheric oxygen (Derry *et al.*, 1992). Thus, the first multicellular eukaryotes would have lacked a circulatory system, and oxygen would have reached inner cells only by diffusion, a process requiring a high external concentration.

The development of a multicellular organism, with differentiated cells, requires that three problems be solved:

Gene regulation. Weismann (1889) believed that different 'ids'—his name for genes—passed to different tissues during development: a leg becomes a leg because its cells contain only leg-making genes. Interestingly, however, he was aware that there is an alternative possible explanation—the one we now think to be correct: with few exceptions, all genes are present in all tissues, and differentiation arises because different genes are active. The first general problem, therefore, is the nature of this gene regulation: how are different genes switched on in different cells?

Cell heredity, and the dual inheritance system. Not only are different genes switched on in different cells, but the states of differentiation are heritable through cell division. In tissue culture, fibroblasts give rise to fibroblasts, epithelial cells to epithelial cells, and so on. When the imaginal discs in larval *Drosophila*, destined to give rise to legs, wings or antennae, are serially transplanted from one larva to another, they also retain their state of determination through many transfers, and many cell divisions (Hadorn, 1965). There is a 'dual inheritance system' (Maynard Smith, 1990). The primary system, depending on nucleotide sequence, determines resemblance between parents and offspring: the secondary system requires the transmission, through cell division, of states of gene activity. How does this secondary system work?

Spatial patterns. Differentiated cells are arranged in space in a specific and repeatable pattern. How does this come about? There are two possibilities, which we will call 'self-organization' and 'external specification'. The latter process is easier to understand: two cells, A and B, become different because of some external influence. For example, in the embryonic induction of the neural tube in vertebrates, typical ectodermal cells differentiate into neural tube because of contact with the underlying embryonic notocord. Another example of such external influence is discussed in Chapter 14: the difference between the anterior and posterior ends of the developing *Drosophila* egg are caused by external stimuli from nurse cells in the maternal ovary, at the anterior and posterior poles of the egg. It may be that this is all there is to

spatial patterning: a succession of external stimuli, spreading back through the generations into the remote past. An alternative kind of process, self-organization, was first clearly formulated by Turing (1952). A sheet of initially identical cells may become different, not because of any external stimulus but because the initial spatially homogeneous state is unstable, and spontaneously generates a spatial pattern: Brownian motion would be sufficient to break the initial symmetry. Are such self-organizing processes important in development? If so, can they generate more than a single difference—for example, between anterior and posterior? Do they give rise to repeated patterns by the generation of standing waves of chemical concentration?

In this chapter, we describe developmental processes that occur in simple organisms: that is, organisms other than animals, plants or fungi. In Chapter 13, we describe the processes of cell regulation and heredity in more detail. We conclude that the cellular processes employed in the development of higher organisms also occur in protists, and often in prokaryotes. It is hard to see what cellular processes, if any, had to be invented before the Cambrian explosion could occur.

It is not sufficient, however, that there be mechanisms for switching genes on and off, and for transmitting states of differentiation through cell division. It is also necessary that differentiated cells be arranged in the appropriate three-dimensional structures. Some protists, and even prokaryotes, form such multi-cellular structures: these are briefly described in this chapter. In Chapter 14, we describe the development of spatial patterns in animals and plants.

Chapters 12–14, then, summarize what is now known about the mechanisms of development. They are concerned with evolution only in so far as they attempt to show that the cellular processes needed for development are present in rudimentary form in existing protists, and hence were probably present before the Cambrian explosion. The additional processes needed for the development of three-dimensional form, described in Chapter 14, are fascinating, but they need not have been difficult to evolve. The fact that such processes have evolved independently on at least three occasions confirms that they can evolve rather readily in response to ecological opportunity.

12.2 The limits of self-assembly

Phage morphogenesis occurs by self-assembly (Darnell *et al.*, 1986). The molecules and supramolecular components, owing to their complementary surfaces, fit together like the pieces of a jigsaw puzzle. The essential feature of such self-assembly is that the shape of the final structure is determined by the shapes of the component molecules. (We later discuss 'self-organization': this is quite

different from self-assembly.) Self-assembly is also in part responsible for the morphogenesis of cell organelles. The process is in no way mysterious. Thus, we know how DNA specifies the amino acid sequence of proteins, and we know the principles on which this primary structure determines the 3-D structure of the protein, even if computational difficulties prevent us from predicting it. The step from the 3-D structure of proteins to the form of a virus, or an organelle, also presents no conceptual difficulty. It is therefore tempting to suggest that cell morphogenesis, too, is akin to crystallization, resulting in stable conservative structures at thermodynamic equilibrium, rather than dissipative structures requiring a continuing input of energy for their maintenance.

A simple example showing that this cannot be so is afforded by cell-wall formation in many unicells. The shape of such cells is set by that of the wall, but the wall itself is not self-assembled from the pieces of some molecular Lego set. Rather, it is the result of the temporally controlled action of enzyme molecules. Even though the end result is very close to a conservative structure, the assembly process needs a continuous input of metabolic energy. The spatial scale of cellular morphogenesis is in the range 1–100 μm, far beyond the effective distance of intermolecular forces. Cellular morphogenesis is an exercise in non-equilibrium thermodynamics and cell physiology, rather than self-assembly.

The crucial force in the morphogenesis of walled cells is very often the minimization of surface stress caused by the inside turgor pressure in these organisms. Synthesis of the cell wall is accomplished by localized cuts in the preexisting fabric of wall molecules, local expansion and re-sealing of the puncture. In prokaryotes, this is carried out in the complete absence of mechanical molecules (the only real mechanochemical devices in prokaryotes are the ribosome 'tape-recorder' and the flagellar motor). In eukaryotes, apical growth of fungal hyphae, budding and cell division of yeasts, and germination of fucoid embryos (see section 14.2) are made possible by the active participation of the cytoskeleton, especially the actomyosin complex.

Simple macromolecular structures may arise by self-assembly, but larger-scale ones usually depend on the sequential activation of genes. There are, however, some developmental sequences that occur too rapidly to be explained by sequential gene activation. An extreme example is the germination of the zoospore of *Blastocladiella,* which consists of the following steps: (1) reception of an environmental signal; (2) the swimming zoospore settles down and retracts its cilium; (3) the nuclear cap breaks down and releases ribosomes; (4) the mitochondrion is fragmented and gamma particles are activated; and (5) a primary cell wall is laid down and the germ-tube initial emerges. All this takes place within a few minutes, without transcription of novel genes.

We conclude that self-assembly plays a role in cell morphogenesis—for example, in the formation of ribosomes—but that larger-scale structures require the differential activation of genes, in time and space.

12.3 The organization of gene action in time: the cell cycle

Development requires controlled sequences of events. States must follow one another in progression: A → B → C → D → ... → Z. Intermediates in this sequence cannot be skipped: if they are, a teratological malformation is the result. This implies that there must be some means to monitor the completion of one step before the system moves to the next. The cell cycle is a nice illustration of how this problem can be solved (reviewed by Murray, 1992). The system has two basic components: a 'cell-cycle engine', and a set of feedback controls. The engine is a biochemical oscillator that can tick autonomously if left alone (although, as we shall see, usually it is not). The states of this biochemical clock trigger morphological events, as shown in Fig. 12.1.

One crucial task is to coordinate the completion of DNA replication with the onset of mitosis, ensuring that they do not get out of phase. There are two known solutions to this problem, in different cell types: relative timing and feedback/checkpoint control. 'Relative timing' means that the rates of different processes are independently regulated so that events occur in the right order. This requires a reliable setting of the rates of DNA replication and MPF—'mitosis promoting factor'—activation, ensuring that the former is completed before the latter begins. If, by experimental manipulation, DNA synthesis is blocked, these cells undergo mitotic division without DNA replication, and the result is death.

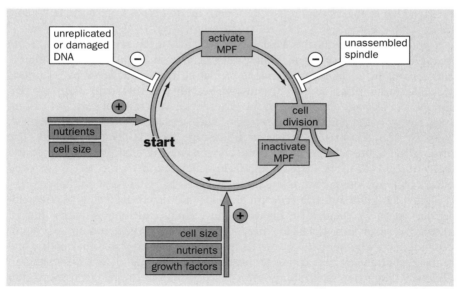

Fig. 12.1 The cell-cycle engine. Positive and negative controls regulate the mitosis-promoting factor (MPF).

The second mechanism is immune to such manipulation: activation of MPF is inhibited by unreplicated DNA, so that blocked DNA replication results in cell-cycle arrest, until the block is removed.

12.4 The 'development' of a unicellular organism: budding yeast

Although budding yeast exists as a single cell, it has several cell types in its life history. Not only are different genes active at different times, as in the cell cycle, but states of differentiation, characterized by different gene activities, are transmitted through cell division. The regulatory systems responsible for the determination of cell type are highly suggestive of analogous processes in multicellular eukaryotes. In fact, the relationship is often closer than this, and amounts to homology. The following aspects, all important in developmental biology, will be discussed: differential gene expression; external induction of differentiation; regulatory hierarchy; and combinatorial control.

The life cycle of *Saccharomyces cerevisiae* is relatively well known (for a review, see Herskowitz, 1988). It consists of a regular alternation of haploid and diploid phases. There are two haploid mating types, a and α. A cell of one mating type can mate only with a partner of the other, to form an a/α diploid cell. Diploids can form spores through meiosis. We are primarily interested in homothallic yeast, which is able to generate cells of both mating types from a single haploid cell.

a and α are alleles of the *MAT* locus, which is a top regulator gene in the hierarchy. Although only one of these alleles can be active at a time, the other is also present in a silent form. Change of mating type is achieved by a curious 'cassette' mechanism: on the chromosome bearing the *MAT* locus, there are two storage loci as well, one for a and the other for α. When a cell switches mating type, a copy of the previously unexpressed allele replaces the one at the playback locus (*MAT*). The original allele at the storage locus is always retained. The mechanism is intrachromosomal gene conversion, initiated by a double-strand break in the *MAT* locus. Only those cells in which another gene, *HO*, is switched on are able to change mating type: the *HO* gene is switched off in diploids.

Figure 12.2 illustrates the control of genes that determine cell type. Haploid-specific genes (*hsg*) must be turned off in diploids, and meiosis must not be allowed to occur in haploids. Further, different mating type genes are expressed, depending on whether *MAT* or *MAT*α is active. It is remarkable that both a and α alleles contain a region similar to the homeodomain that we will meet again in Chapter 14 in genes determining segment specificities in *Drosophila*. It seems that the homeodomain played a role in controlling the activities of other genes before the origin of multicellularity.

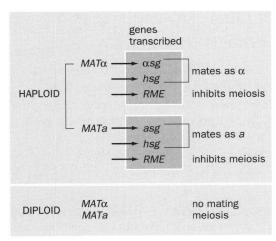

Fig. 12.2 Control of genes specific for cell type in budding yeast. In haploid cells, the 'play-back' locus is occupied by one of two alleles, *MATα* or *MATa*. These genes activate mating-type specific genes, *α* or *a*, and also haploid-specific genes, *hsg*, and a gene *RME* that inhibits meiosis. In the diploid, none of these regulated genes is active. The control system involves transcription factors with positive and negative effects on transcription: some transcription factors are formed by the union of products from two different genes.

We now turn to the induction of differentiation. Although the *a* and *α* alleles are responsible for the differences between the corresponding cell types, the respective phenotypes appear in full only when cells of opposite mating type come into contact. *α* and *a* cells code for *α* and *a* pheromones, respectively; *a* and *α* cells have receptors only for the pheromone produced by cells of the opposite mating type. When *α* pheromones are bound by the receptor of an *a* cell, they activate the *a*-specific genes. Thus, the pheromones induce hyperexpression of genes specific for the opposite mating type. This demonstrates a recurrent theme in developmental biology: the differentiated state depends on intrinsic programming and on response to the environment.

Note that there is a regulatory hierarchy and a sequential activation scheme involved in these processes. At the top of the hierarchy are those genes that regulate the expression of *HO*. Activation of the latter depends on three conditions: the haploid state (cell-type control), late G1 phase (cell-cycle control) and the cell being a mother cell (asymmetric control). The mating-type switch affects the *MAT* locus, which in turn is responsible for the maintenance of the specialized cell types. We will meet a similar regulatory hierarchy in the early development of *Drosophila*. Note that the cassette mechanism is a peculiarity of mating-type switching in yeast: the job could be done without chromosomal rearrangement.

A final complication is that control is not always exerted by a protein specified by a single gene: complexes formed by two or more peptides may act as transcription factors. Such combinatorial control is widespread in multicellular organisms, but is already present in a simple form in yeast.

The one developmental process than cannot in principle be exemplified in single-celled organisms is that whereby differentiated cells become arranged in a spatial pattern: we now turn to some simple examples of this process.

12.5 The division of labour in the origin of multicellular eukaryotes: *Volvox*

Adam Smith, in *The Wealth of Nations* (1776), discussed the importance and conditions of the division of labour in society. Three principles emerged:

- Division of labour is worth while if there is increased dexterity resulting from concentration on a single task, the omission of the need for switching between tasks, and the relative ease of inventing specialized machines.
- Division of labour pays only if the market is sufficiently large.
- In markets of a given size, the degree of division of labour will depend on the ease of communication.

Bell (1985) considered analogous problems in the organization of multi-cellular organisms, illustrated by the various species of the Volvocales (see also Bell & Koufopanou, 1991: for the basics of *Volvox* life cycles, see Kirk & Harper, 1986; Kirk, 1988).

The specific name *Volvox weismannia* is a reminder that these algae are an excellent illustration of the stepwise appearance of the segregation of germ line and soma, first emphasized by Weismann (1889). He also suggested that *Volvox* might provide particularly favourable material for investigating the origins of the dichotomy.

Volvox is closely related to the unicellular alga *Chlamydomonas*, in the order Volvocales. Most members of the order possess only a single cell type that fulfills all vegetative and reproductive functions. In the genus *Pleodorina* there is a temporal division of labour: some of their cells start with vegetative functions, but later differentiate into gonidia (the asexual propagules). The genus *Volvox* has a bona fide segregation of germ line and soma: germ cells are immotile in the interior of the spheroid, and somatic cells bear cilia but cannot divide. Figure 12.3 illustrates the asexual life cycle of *Volvox carteri* (originally named *Volvox weismannia*).

The process of cleavage and inversion of gonidia forms a fascinating parallel to gastrulation in metazoans. Juvenile gonidia leave the parental spheroid by making holes in its surface: the latter then undergoes senescence and pro-grammed cell death. Cleavage of the gonidium results in a morula-like ball of cells. Asymmetric divisions quickly result in the formation of gonidial precursors, in a precise temporal and spatial arrangement. At the end of cleavage, cells are located only on the surface of the sphere. This is a blastula-like stage, the snag being that it is completely inside out: gonidia, which are giant cells compared to the ciliated somatic cells, are anchored to the outside of the sphere. This situa-tion is corrected by inversion, starting at the so-called phialopore, a hole in the sphere. After successful cleavage, a wave of contraction sweeps over the embryo, the phialopore opens wide, and the four lips of cells adjacent to the phialopore

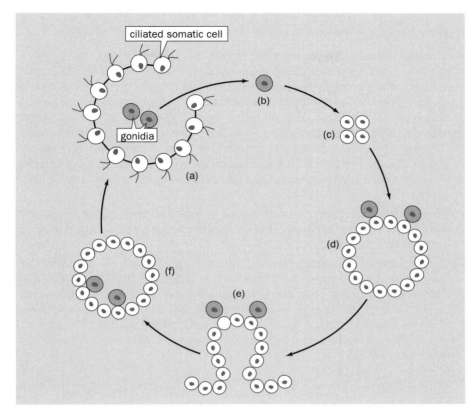

Fig. 12.3. The asexual life history of the alga *Volvox*. (a) Adult spheroid of ciliated somatic cells, containing gonidia. (b). The gonidia escape from the spheroid, which dies. The gonidial cell cleaves (c) to form a ball of somatic cells, with newly differentiated gonidial cells attached to its outside (d). (d–f) A process of inversion, analogous to gastrulation, brings the gonidial cells to the interior of the spheroid.

curl backwards over the surface of the embryo. The zone of curvature accompanying this backward bending of the lips now moves in a wavelike fashion, involving cells further and further from the phialopore, and bringing more and more of the interior hemisphere to the exterior. Suddenly, when the region of maximum curvature has passed the equator, the posterior hemisphere 'pops through' the opening. After this inversion—or rather, as it should be called, eversion—the phialopore lips are brought back together.

The driving force behind these radical movements is the change in shape of individual cells, achieved through programmed action of the cytoskeleton. The neighbours of each cell remain invariant throughout. The connections between cells are made by cytoplasmic bridges: in this sense, *Volvox carteri* is a syncitium.

Following Bell (1985), we now ask how far Adam Smith's ideas on the division of labour are relevant to the evolution and development of *Volvox*.

- All members of the Volvocales must be motile and reproductive. In *V. carteri*, this is achieved by a complete division of labour. There is a striking difference in both size and morphology between somatic and vegetative cells. The benefit of this differentiation has been demonstrated by comparing the growth rates of unicells and colonies. For the same overall size, growth rate is always higher if there is division of labour. Colonies produce a greater bulk of smaller offspring than do unicells of comparable size. A possible mechanistic explanation is that gonidia are specialized for synthetic and mitotic activity, whereas somatic cells are specialized for energy output—in particular, ciliary activity.

- The size difference between small vegetative cells and large gonidia increases with colony volume. Division of labour does not occur below a critical volume in the Volvocales.

- The importance of communication (and transport) is nicely illustrated by comparing *Merillisphaera* and *Euvolvox*. In the former, the cytoplasmic connections between cells are lost during inversion, whereas they persist through the whole life cycle in the latter (it is for this reason that the cells of *Euvolvox*, viewed from above the surface, resemble little starfish). The growth of reproductive tissue, relative to that of the somatic cells, is much greater in *Euvolvox*.

12.6 Multicellularity through aggregation: myxobacteria and slime moulds

Multicellularity is not confined to the eukaryotes. Myxobacteria are Gram-negative, multicellular prokaryotes, whose sporulation is preceded by aggregation and the development of sometimes quite elaborate fruiting bodies. The life cycle shows an astonishing similarity to the better known slime mould development (Fig. 12.4). The similarities are as follows (Kaiser, 1986):

- Haploidy, and asexuality of the spores formed in the fruiting bodies.
- A common habitat—wet soil.
- Movement on a surface: neither can swim.
- Starvation stops cell reproduction and initiates development: amino acid starvation is essential for such initiation.
- During aggregation, travelling waves can be observed.
- Both make molecules with CAM-like activity during activation.

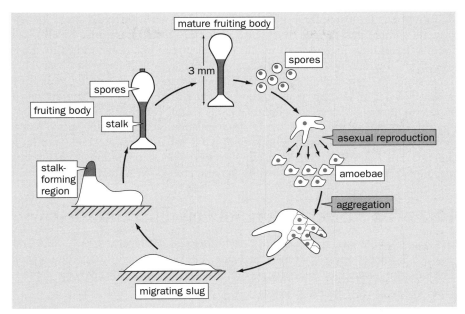

Fig. 12.4 Life history of the slime mould *Dictyostelium*. The mature fruiting body, 3 mm high, produces spores. These hatch into amoebae, which multiply asexually. Aggregating amoebae form a migrating slug, which develops into a fruiting body.

- Not all cells become spores: some lyse (*Myxococcus*) or become dead stalk cells (*Dictyostelium*).

- The fruiting bodies contain 10^5 to 10^6 cells.

In some slime moulds, intercellular signalling happens by the release and detection of cAMP in the environment. The release and decomposition of cAMP is an oscillatory process, resulting in the coordinated aggregation of the amoeboe (for a detailed explanation, see Darnell *et al.*, 1986). Some functionally similar signal molecule must exist also in myxobacteria, since cells that are already aggregating can trigger others to do so without close cell-to-cell contact.

What is the selective advantage of such a life cycle? In myxobacteria, it may have started through cooperative feeding and growth (Shimkets, 1990). Myxobacteria are predators of other bacteria: they secrete digestive enzymes into the environment, and feed on the digested material of the attacked bacteria. Since, due to the lack of a cytoskeleton, they cannot feed by phagocytosis, digestion is more effective if it occurs in a space surrounded by myxobacteria. The common advantage of a fruiting body must lie in more efficient spore dispersal (Bonner, 1982). The differentiation of spore and non-spore cells is a good example of the division of labour, provided that mechanical and spore-formation efficiency are partly in conflict.

Since these multicellular organisms are formed anew in each cycle by aggregation, there is a danger that selfish cells (e.g. refusing to form a stalk) will disrupt the multicellular organization. The selective force preventing this must arise from population structure: gliding bacteria, as well as crawling amoeboe, move slowly. Given limited movement, the spores in a single fruiting body may be descended from a single spore in the last generation, or a recent generation. If so, the maintenance of multicellularity is, as usual, partly due to kin selection. But this cannot be the whole story, since cheaters will inevitably arise: when that happens, they are presumably eliminated because aggregates containing cheaters are less efficient in dispersing spores. This explanation is at present speculative, and should be checked by genetic studies on natural populations.

12.7 Two mechanisms of cell differentiation

One can imagine two models of the early Metazoa (Wolpert, 1990). The first (Fig. 12.5a) depends on differences between cells that arise at the moment of cell division. The propagule divides by asymmetric cell division. The two cells remain

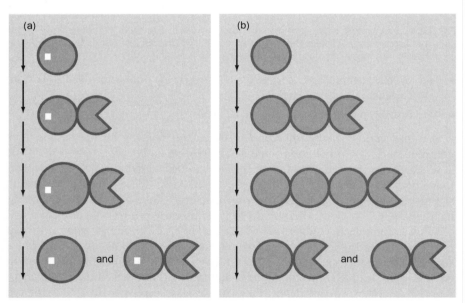

Fig. 12.5 Two models of cell differentiation (after Wolpert, 1990). The organism is made up of just two cell types, one of which can be thought of as specialized for feeding. In (a), development depends on lineage, and the existence of a stem cell: the white square represents a determinant that maintains the non-feeding cell as a stem cell. In (b), development depends on cell-to-cell interactions: this requires both cell signalling and polarity.

attached to one another. One retains the capacity to divide, whereas the other is specifically adapted to feeding, at the expense of losing the ability to divide. When the 'stem cell' divides again, one of the daughter cells remains attached to the feeding cell, and the other must divide again to produce a new feeding cell. This illustrates a primitive differentiation between germ line and soma, with an associated division of labour. For the process to work, a stem cell must 'know' whether it is attached to a feeding cell or not. This does not require cell-to-cell communication: it is sufficient that there be two kinds of cell division, one producing an attached feeding cell, and the other to an unattached stem cell. This requires a regulatory gene which periodically switches between two states, triggered by cell division. Detailed genetic networks of such 'toy models' of development can be found in Szathmáry (1994). Such an organisn would be incapable of regeneration: 'knowledge' about the state of differentiation is pre-programmed, and does not depend on checking the state of neighbouring cells.

An alternative process (Fig. 12.5b) relies on cell-to-cell interactions. An imaginary two-celled organism reproduces by binary fission (vegetative re-production), requiring two cell divisions. After the second division, the two central cells become decoupled, which triggers regulative differentiation of a new feeding cell in the offspring lacking one. Both cell polarity and cell-to-cell communication are needed: cells must be informed about the state of neighbouring cells. Such organisms are expected to show regeneration upon injury. Although the model is imaginary, it is interesting that the very simple placozoan, *Trichoplax*, has only two cell types, and reproduces by fission.

The point of these models is that they illustrate in the simplest form the distinction between mosaic and regulative development: both types of process occur in higher organisms.

13 Gene Regulation and Cell Heredity

Two cellular mechanisms are essential for development. The first, gene regulation, makes it possible to switch on different genes in different cells, in response either to conditions external to the cell or to the activity of other genes within the cell. The second, cell heredity, ensures that these states of gene activity, once induced, can be stably transmitted through cell division, without the need for the continued presence of an external inducer. In this chapter, we describe how gene regulation and cell heredity are achieved in metazoans, and point to some similar mechanisms that are already present in prokaryotes.

13.1 Gene regulation

The central problem of gene regulation was posed, in a social context, by the scholastic Master Eckhardt: 'Quis custodiet ipsos custodes?' [Who regulates the regulators?] Clearly, the proposition that every gene needs a separate regulator gene leads to an infinite regress. There are various ways of resolving the paradox, which include (1) one regulator controls several other genes, including regulators; (2) one gene, even a regulator, is controlled by several other genes; and (3) some genes may be both regulatory and structural. Plenty of examples are known for each case. It is also necessary that some genes be regulated by signals from outside the cell.

The essential mechanism of gene regulation was discovered by Jacob & Monod (1961; Fig. 13.1) in *E. coli*. A regulatory gene codes for a protein, which, by binding to a specific regulatory sequence of another gene, alters the activity of that gene (negatively in the case originally described by Jacob & Monod, but the effect can also be positive). The regulation can be modified by a specific inducing molecule that alters the effect of the regulatory protein by binding to it allosterically. It is interesting that these two properties of regulatory proteins—that they can recognize specific regulatory sequences, and that their effectiveness can be altered by binding allosterically to inducers—are already present in prokaryotes.

The complexity of multicellular eukaryotic development requires that an average gene be controlled by many others. Whereas regulatory elements in bacteria are usually simple switches, eukaryotes tend to have 'smart' genes, controlled by a complex of several regulatory proteins (Davidson, 1990; Beardsley, 1991). Eukaryotic genes have so-called enhancer elements associated with them that may not be close on the chromosome at all: some transcription factors bind 40 000 base pairs from the target gene. The average number of regulatory sites for eukaryotic genes is estimated to be around five, in contrast to the maximal number of three in prokaryotes. One actin gene in a sea urchin has as many as 20 regulatory sequences.

Davidson (1990) pictured a smart histospecific gene as having regulators for time, histospecificity, lineage or regional identification, cell-cycle linkage, ne-

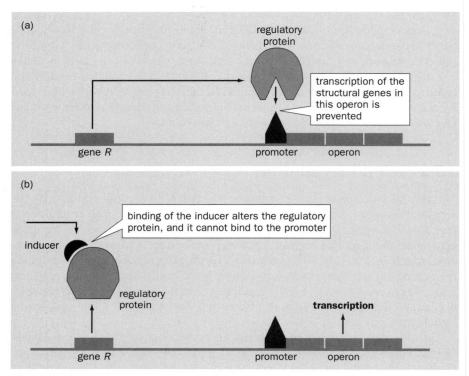

Fig. 13.1 Regulation of the *lac* operon (Jacob & Monod, 1961). (a) A gene, *R*, produces a regulatory protein, which binds to a promoter, thus preventing transcription of the structural genes in the operon. (b) An inducer binds to the repressor protein, altering it so that it cannot bind to the promoter, thus permitting transcription of the operon.

gative function, amplitude control and reception of signals transduced from adjacent cells. As we saw from the examples of yeast cell types, and will meet again when describing pattern formation in *Drosophila*, genes belong to regulatory hierarchies. For example, the gene *engrailed* is active in seven stripes in the early *Drosophila* embryo, but it seems that in each segment it is switched on by a different but unique combination of regulatory genes.

13.2 Cell heredity

It is characteristic of cell heredity that acquired characters are inherited. The experimental protocol is as follows: cells are incubated in an environment that precludes cell division, and hence the possibility of selection, and are later tested for the establishment of novel traits. Landmann (1991), in an interesting review,

says: 'Following the exposure, and upon being returned to the original environment, all or a large proportion of the treated cells (or organisms) exhibit new characteristics that are passed on heritably to succeeding generations.'

An excellent prokaryotic example is the heritable induced state for β-galactosidase synthesis in *E. coli*—another aspect of the famous operon model. At a very low concentration of the inducer, there is no appreciable enzymatic activity. If the cells are exposed to high inducer concentration, they will, of course, become induced, and show a high enzymatic activity. The surprising fact is that if they are put back into the first medium with a low concentration of inducer, the cells remain in the induced state, at least for 180 generations.

The explanation is as follows: in the inducing environment, high levels of β-galactoside permease, as well as β-galactosidase, were induced. When the cells are returned to the low-concentration solution, the permease is still active, and keeps on pumping the inducer into the cell. Thus, the inducer concentration inside the cells is higher than in the solution, and sufficient to maintain the induced state. Note that, even though there is no change in DNA sequence, an element of autocatalysis is essential: it is the transcription activity of the operon genes that can be said to be autocatalytic.

In eukaryotic development, however, a more important mechanism of cell heredity is the copying of some mark or label on the DNA at the time of replication. This also is not without prokaryotic precedent: the host-induced modification of bacteriophage T2. This depends on the restriction-modification system of bacteria, which can label their own DNA with specific glucosylation or methylation markings, and cut into pieces, with the aid of restriction endonucleases, foreign DNA that is not so labelled. T2 phages grow vigorously in *E. coli* cells whose restriction-modification system is intact. When phages from the same culture infect a bacterial strain that is deficient both in modification and restriction, a single round of phage replication is sufficient to render the phages incapable of growing in normal bacteria, although they continue to be viable in the deficient mutants.

The reason is that phage DNA is normally modified on both strands. New labels are attached at replication to the daughter strands. In the deficient bacteria, the labelling of new strands does not occur, so that the phage become prone to restriction cutting in the wild-type bacteria. Because the mutant bacteria are also lacking the restriction system, unlabelled viruses can grow in them.

The process of methylation can give rise to phenotypically different strains of bacteria, which remain different through many cell divisions, although genetically identical. For example, *E. coli* cells can have various types of pili, which attach them to various host tissues. The production of pili is not constitutive, but depends on appropriate environmental conditions. Even in favourable conditions, however, pili may be absent: the cells can be in one of two 'phases', ON and OFF (Van der Woude *et al.*, 1992). The difference is stably inherited through cell division. It depends on which of two GATC sites upstream of the gene is

methylated, and also on specific methylating enzymes, and transcription factors that bind to these sites. In other words, prokaryotes also possess a second inheritance system, depending on the labelling of DNA rather than on changes in sequence, analogous to the system responsible for cell differentiation in multicellular organisms.

Similar labelling takes place in at least some eukaryotes. It has been known for some time that methylation of cytosine at CpG dinucleotides in DNA affects transcription of mammalian genes. Methylation alters local chromatin structure, which in turn affects accessibility to regulatory proteins and the endonuclease DNAse I. The general rule is that inactive genes (e.g. tissue-specific genes in the germ line) are methylated, and become active when they are demethylated.

This system can be fooled with the cytidine analogue 5-azacytidine, which can replace C in DNA but is an inhibitor of DNA transmethylase. Hence, upon treatment with this drug, cells are expected to lose some of their methylation marks, and some of their genes should be activated. This prediction was confirmed when a mouse embryonic cell line, with a fibroblast morphology, was transiently treated with the drug. Cells in 1500 colonies were examined for altered morphology. The newly apparent phenotypes included adipocytes (for fat storage), typical myocytes and chondrocytes. Controls did not contain such differentiated cells. It was commitment, rather than differentiation, that was caused by the treatment: some cells that did not show signs of differentiation produced differentiated offspring later. The new phenotypes were stably inherited for several hundred generations.

All this makes good sense from an embryological point of view. There must be some means to switch between cell types, but those differentiated cell types that reproduce by cell division must be able to pass their identity to their daughter cells: liver cells are generated only during development, but a liver cell must give rise to a liver cell.

Higher plants and mammals show extensive methylation patterns. Methylation and demethylation of gene promoters is correlated with inhibition and activation of transcription, respectively. Although, in most cases, it is not clear whether methylation is a cause or an effect of gene activation, there are some experiments, including the one described above, indicating that at least sometimes it is the methylation state itself that determines gene activity. In all cases examined, methylation patterns are stably propagated: this maintenance of methylation by a transmethylase is not sequence specific, but depends only on the semiconservative nature of DNA replication. Clearly, the cell must have some means of altering methylation patterns. The CpG dinucleotide by itself cannot provide specificity, so adjacent sequences must participate in the labelling–delabelling processes, which are presumably carried out by enzymes that recognize the relevant sequences.

Although some of the clearest examples of cell heredity come from *Drosophila*—for example, the propogation of imaginal discs mentioned above—DNA is

not methylated in this organism. We assume that there is some other label on the DNA and this is replicated when the DNA is replicated, but at present this is mere supposition.

13.3 What had to be invented?

The main motive for our discussion of gene regulation and cell heredity was a wish to answer the question: what had to be invented before the evolution of the Metazoa became possible? We have to admit that no clear answer has emerged. It is possible to see why the evolution of multicellular prokaryotes has been limited, although it has certainly occurred. Bacteria already have mechanisms for gene regulation, and, in the host-induced modification system, they have a way of labelling specific DNA sequences, and of transmitting the resulting labels through cell division. Thus, the capacity for gene regulation and for cell heredity are present. Further evolution of multicellularity may have been limited by the following constraints:

- Reduced motility in the absence of mechanoproteins. It may be that, without a cytoskeleton, cell-wall growth cannot be modified to make filament branching possible: there are no bacteria with branching.

- Limited means of development. Movements requiring cytoskeletal activity, such as gastrulation, are excluded. Animal-like organisms, able to ingest prey, are excluded altogether.

- Genome size and genetic complexity are limited in prokaryotes by the single origin of chromosome replication.

- Limited means of chemical communication with the environment: lack of the Golgi apparatus.

- Metabolic limitations. There is no proper separation of the respiratory chain and the photosynthetic apparatus in bacteria that have both: they must be inferior to plant cells. Bacteria that only respire are inferior to animal cells since they cannot phagocytose. Only simple fungus-like organisms are possible without branching filaments.

Finally, metabolic efficiency seems to be a rather strong precondition for multicellularity. There are very few anaerobic multicellular eukaryotes, although some fungi have secondarily lost their mitochondria. There are no multicellular archaezoans (eukaryotes without mitochondria).

If it is hard to see what prevented prokaryotes from evolving complex multicellular forms, it is still harder to see what the eukaryotes were waiting for. If *Volvox* can gastrulate, and segregate its germ line from its soma, what remains? (Of course, *Volvox* is a recent organism: perhaps no such inventive algae existed in the Precambrian.) One is left with the impression that what the eukaryotes

were waiting for was the ecological opportunity, provided, perhaps, by an increase in atmospheric oxygen. The main fact pointing in the opposite direction is the apparent monophyletic origin of the Metazoa: if there were no intrinsic constraints on the evolution of multicellularity, one would expect multicellular clades to have originated independently from several protist ancestors. But this is not a decisive argument. After reaching the conclusion that it was ecological opportunity rather than any cellular invention that triggered the Cambrian explosion, we found that Wolpert (1990) had earlier concluded that 'evolution of development required few new features not already present in the eukaryotic cell, as exemplified by the cell cycle'.

How can we expect to gain further insight? The best hope lies in molecular genetics. Molecular data are the main reason for accepting the monophyletic origin of metazoans. The homeobox story, discussed in Chapter 15, tells us that the common ancestor of the chordates and the arthropods probably already had a differentiated head, middle and tail. The regulatory genes responsible for the differentiation were already present. Since their DNA-binding region, the homeobox, is also present in the gene-determining mating type in yeast, it is likely that the common ancestor of these genes already existed in their protist ancestor. No-one could have foreseen these observations. We can reasonably hope that future discoveries will be equally illuminating.

14 The Development of Spatial Patterns

14.1 Flower development as an example of morphogenesis

It is convenient to start with an account of the development of the flower in the crucifer *Arabidopsis* (Coen & Meyerowitz, 1991), for several reasons. The development of plants is in one respect simpler than that of animals: cells do not move relative to one another. Much of animal development is achieved by cell movement, contraction and adhesion; these are processes we need not bother with when thinking about plants. As yet, less is known of *Arabidopsis* than of *Drosophila* or of vertebrates: what is known is simple and elegant, and brings into sharp focus the things that are not yet known.

The flower develops from a disc of cells (Fig. 14.1). Four concentric rings differentiate. The outermost ring gives rise to four sepals, the next innermost ring to four petals, the next to six stamens, and the central cells to two fused carpels. Mutants are known that alter this pattern. Usually, a given mutant alters two neighbouring rings of organs (Table 14.1). A simple model explains these mutants (Fig. 14.2). Genes are expressed in three regions, A, B and C. If we symbolize the corresponding genes by *a*, *b* and *c*, then *a* gives rise to sepals, *ab* to petals, *bc* to stamens and *c* to carpels. This would account for normal development, but if we are also to explain the mutants, we must assume that genes *a* and *c* are mutually inhibitory, so that when *a* is absent or deficient, *c* is expressed over the whole region, and vice versa.

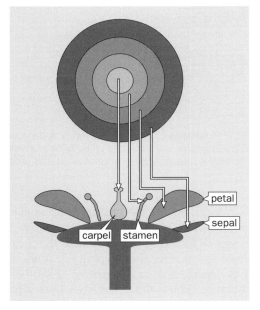

Fig. 14.1 The flower of *Arabidopsis*. Above is the disc of cells from which the flower develops, and below is a section through the flower, showing the structures that develop from each ring of cells.

Table 14.1 Mutants of *Arabidopsis*

Wild type	Sepal	Petal	Stamen	Carpel
Class 1 (*a*)	Carpel	Stamen	Stamen	Carpel
Class 2 (*b*)	Sepal	Sepal	Carpel	Carpel
Class 3 (*c*)	Sepal	Petal	Petal	Sepal

Table 14.2 Double mutants of *Arabidopsis*

Wild type	Sepal	Petal	Stamen	Carpel
b⁻ c⁻	Sepal	Sepal	Sepal	Sepal
a⁻ b⁻	Carpel	Carpel	Carpel	Carpel
a⁻ c⁻	Leaf	p–s*	p–s*	Leaf
a⁻ b⁻ c⁻	Leaf	Leaf	Leaf	Leaf

* p–s indicates a structure intermediate between a petal and a stamen.

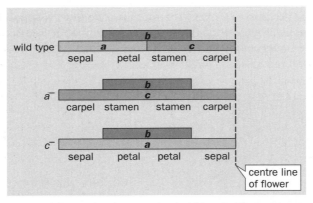

Fig. 14.2 A model of flower development in *Arabidopsis*. Three genes, *a*, *b* and *c*, are expressed in different regions of the disc. They determine floral structures as follows: *a* only, sepal; *a* + *b*, petal; *b* + *c*, stamen; *c* only, carpel. If gene *a* or *c* is inactivated, the other gene is expressed across the whole disc.

This model has been confirmed in two ways. First, it can satisfactorily explain double mutants (Table 14.2). Triple mutants, in which all three gene activities are missing, develop leaf-like structures in all four rings: this confirms the long-held view that flower organs are modified leaves. The phenotypes of the various double mutants can be predicted from the model, and are as expected. The model has been confirmed more directly by hybridization *in situ* of the RNA transcripts

of the different genes, which appear early in the development of the flower primordium in the expected places.

An almost identical system is present in *Antirrhinum*. There is molecular homology between the genes in the two species. For example, the class 3 mutant *agamous* of *Arabidopsis* is homologous to the mutant *deficiens* of *Antirrhinum*. The two genes have a common DNA-binding region of 56 amino acids, of which 40 are identical. A similar region, called the MADS box, is present in some of the other genes.

Two questions at once arise:

- What causes class 1 genes to be switched on in the outer ring, class 3 genes in the centre, and class 2 genes in the intermediate region? The short answer is that we do not know. A plausible guess is that, before the genes are switched on, a gradient of some substance or substances arises from the centre to the outside of the primordium, and that the varying concentrations of this substance activate the genes. In Wolpert's (1969) terminology, the gradient provides 'positional information'. In section 14.3, we discuss how such information is provided in some animal examples.

- If positional information is present, why are these genes necessary? Why should not the substance whose concentration provides positional information directly switch on all the genes whose products are needed in flower growth? Why first switch on genes *a*, *b* and *c*, and then use them to switch on other genes? The answer probably lies in the need for accuracy. For example, it is interesting that genes *a* and *c* are mutually inhibitory, so that only one of them is active in any particular cell. Consequently, only petal-making genes or stamen-making genes, but not both (or neither), are switched on. If these genes were directly controlled by the chemical gradient (supposing one to exist), it is inevitable that there would be cells in which both types, or neither, were active. The general principle seems to be that only a limited amount of spatial complexity can be gained in a single step, and that complex forms require a cascade of gene activities, hierarchically arranged.

14.2 Positional information: external specification or self-organization?

In *Arabidopsis*, different regulatory genes are switched on in different places, and, by their activity, induce the development of different organs—sepals, petals, stamens and carpels. The unanswered question is how the positions at which the genes are to be active are specified: in Wolpert's (1969) term, how is positional information specified? As we suggested in Chapter 12, there are two possible answers to this question: positional information is imposed from outside, or it

appears intrinsically, by self-organization. We now spell out these alternatives in more detail, and consider some experimental evidence.

In his 1969 paper, Wolpert had in mind that information was initially imposed from outside. Pattern arises in an initially homogeneous tissue in response to a monotonic gradient in concentration of some inducing substance, which it is convenient to call a morphogen. The gradient is initiated by some outside influence: we describe below how the anteroposterior gradient in the *Drosophila* egg in the concentration of the *bicoid* gene product arises because it is produced by maternal nurse cells situated at the anterior pole of the egg. Of course, it is not essential to the model that the gradient be set up maternally, or that it be in the concentration of a gene product made by cells external to the developing tissue. It could arise, for example, because cells at one pole of a tissue are activated by contact with an external inducer. Some patterns require the simultaneous presence of two intersecting gradients. The pattern then develops further because cells respond to the local concentration of morphogen(s): typically, the response will be mediated by activating a regulatory gene, which itself activates other genes lower in the hierarchy.

An interesting question is how far it is possible to have more than a simple on–off response: is it possible for cells to respond in several qualitatively different ways to varying concentrations of the same morphogen? The answer is yes. Driever & Nüsslein-Volhard (1988) have studied the response of the *Drosophila* embryo to varying concentrations of the mRNA transcript of the *bicoid* gene. This gene is active in the maternal nurse cells at the anterior pole of the egg, and mRNA concentration can be manipulated by altering the maternal genotype. The authors show that (at least) three qualitatively different responses exist to different concentrations: at low concentrations, gnathal and thoracic structures appear; at intermediate concentrations, head structures appear; and at the highest concentrations, the most anterior structures (anterior to the segmented head) appear. Green & Smith (1990) obtained similar results for the mouse. They exposed mouse blastomeres to varying concentrations of the protein XTC-MIF, a homologue of activin-A, which is known to induce mesodermal structures. They found two thresholds, separating three different responses: a change in concentration by a factor of 1.5 was sufficient to switch the response from epidermal keratin to muscle. The two series of experiments are similar in suggesting that several (> 2) qualitatively different responses are possible to a single gradient. The detailed mechanisms must differ, however, because the bicoid protein is a DNA-binding transcription factor, whereas XIC-MIF acts extracellularly: remember that early differentiation in *Drosophila* occurs in a syncitium.

It is clearly not plausible that the development of adult form depends solely on local responses to the concentration of a single set of intersecting gradients: the black stripes of a zebra are not induced by a series of discrete values of the concentration of a single morphogen. But it is logically possible that adult form

depends on a series of such stages, each adding to spatial complexity, and forming the starting point of the next stage. Is there, then, any place for self-organizing processes in development?

The characteristic feature of a self-organizing process is that a spatial pattern develops in an initially homogeneous tissue, without need of an asymmetric stimulus. The distinction between such a process and an externally imposed gradient is not always clear-cut. For example, consider the early development of the seaweed *Fucus*, explained in Fig. 14.3.

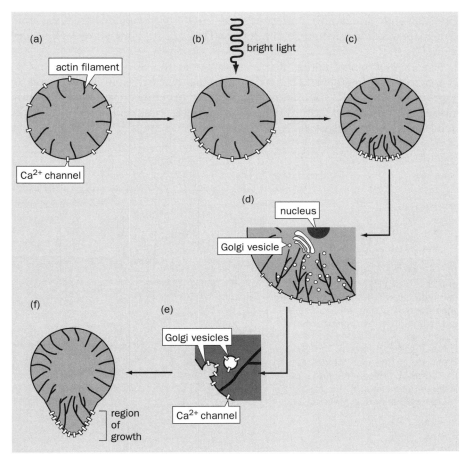

Fig. 14.3 The polarization of the *Fucus* embryo (after Harold, 1990). (a) The initial, symmetrical distribution of Ca^{2+} channels and actin filaments; (b) after illumination, channels migrate to the lower half of the zygote; (c) Ca^{2+} channels initiate growth of a network of actin filaments; (d) Golgi vesicles are guided to the rhizoid pole; (e) vesicles bearing Ca^{2+} channels fuse with the plasma membrane; (f) actin cables extend to the nucleus, and rhizoid outgrowth begins.

Perhaps it does not greatly matter whether one regards *Fucus* as illustrating self-organization or the effects of an external stimulus: in either case, all that is produced is a simple polarity. Self-organization becomes more interesting if it can be shown to generate more complex spatial patterns—in particular, periodic patterns. The theoretical possibility certainly exists. In 1952, Turing published a seminal paper that founded the reaction–diffusion explanation for pattern formation. A simple outline of his hypothesis is as follows (Fig. 14.4). There are two chemical substances, X and Y. It is assumed that X enhances its own formation, by autocatalysis, as well as that of Y (by heterocatalysis). In contrast, Y helps to break down X (inhibition). Both substances are soluble in an indifferent medium. Y diffuses much faster than X. The development of a standing wave is illustrated in the figure.

The revolutionary nature of this work lies in the discovery that diffusion, combined with appropriate chemical reactions, can act as a creative force in pattern formation. It is important that Turing structures are non-equilibrium, dissipative formations: a constant input of energy is required. It is assumed in the model that autocatalysis of X is maintained by a 'nutrient': take away the

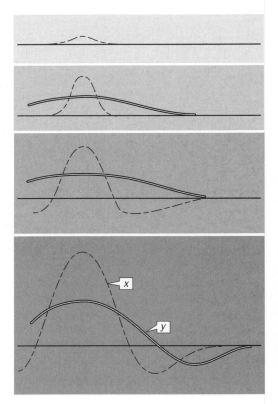

Fig. 14.4 Turing's (1952) theory of chemical morphogenesis. There are two morphogens, X and Y; *x* and *y* are their concentrations, measured as departures from the homogeneous equilibrium. X stimulates its own synthesis, and that of Y. The presence of Y causes the destruction of X. Both morphogens diffuse, but Y diffuses faster than X. The result is the development of a standing wave.

nutrient and the pattern collapses. Note that the model illustrated in Fig. 14.4 is unrealistic in being linear: non-linear analysis, however, confirms the possibility of a standing wave.

Some attempts have been made to apply these ideas to development. Maynard Smith & Sondhi (1961) interpreted the arrangement of bristles in *Drosophila*, including abnormal arrangements in a mutant strain, in terms of Turing waves. Meinhardt (1982) used the model to explain features of *Drosophila* development. Murray (1990), using computer solutions, showed that all known mammalian coat patterns can be explained by appropriately adjusted reaction–diffusion models. Other theoretical developments include the derivation of travelling waves, modified equations for chemotaxis-mediated pattern formation, and mechanochemical models of morphogenesis (reviewed by Murray, 1990). The weakness of all these models is that, although they show that a biological pattern could have arisen by a Turing mechanism, they do not demonstrate that it has in fact done so.

Many biologists remain rather sceptical about such ideas. One reason is that, until recently, no known Turing structure existed, even in the field of chemistry. Some 40 years after Turing's original paper, efforts to produce such a structure have met with success. Castets *et al.* (1990) created a Turing structure in a chemical system consisting of chlorite, iodide, malonic acid and starch. Iodide is an activator and chlorite an inhibitor. The major innovation was the addition of starch, since iodide forms a rather stable complex with it, effectively slowing down its diffusion rate. Previous attempts at implementing chemical Turing systems failed because in water all molecules have similar diffusion rates. Lengyel & Epstein (1991) gave a nice mathematical analysis of the system, and could prove that it was indeed a standing wave. It may be that biological systems, where cell-membrane receptors abound, are more favourable for emerging Turing structures than are pure chemical ones.

The theoretical findings are worth summarizing, particularly concerning the distinction between travelling and standing waves, because it is the latter that we need for embryological induction but the former that are most easily generated experimentally. Standing waves are based on systems with a stable equilibrium when they take place in a spatially homogeneous environment (such as a well-stirred flow reactor): the system must also be unstable to spatial perturbations. In contrast, travelling waves arise when the homogeneous system shows stable oscillations (limit cycles). If such a system is unstable to spatial perturbations, different places will get out of phase, and waves appear, as in the famous Belousov–Zhabotinski reaction.

Perhaps the main reason for the reservations felt by experimentalists towards Turing structures, however, arises from the absence of convincing biological evidence for their existence. The formation of segments in *Drosophila* affords a crucial test case. Looking at Fig. 14.5, showing seven evenly spaced bands, it is hard not to think that a Turing mechanism is responsible. Yet, as explained

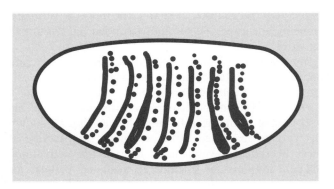

Fig. 14.5 Segment determination in *Drosophila*; rings of cells in which RNA transcribed from the genes *hairy* and *even-skipped* is present.

below, although the same gene is active in each of the bands, there are good reasons to think that it is switched on by a different inducer in each segment. This is not what one would expect of a Turing process. However, the matter is still open. There are other structures—for example, the regularly arranged petals in a flower disc—that may prove to be generated by a wave. And even in the case of *Drosophila* segments, it may be that a Turing wave ensures the regular spacing of the bands, and that the differences between the inducers are present because of the need to specify a different path of later development for each segment. Fortunately, time and further experiments will tell us the answer.

14.3 Positional information in *Drosophila* and the chick

In *Drosophila*, the primary positional information differentiating the anterior from the posterior of the embryo is provided maternally (for reviews of *Drosophila* development, see Ingham, 1988; Lawrence, 1992). As the egg grows in the ovary, substances are manufactured in the maternal nurse cells, and then enter the egg. A maternal gene, *bicoid*, is active in the nurse cells at the anterior pole of the egg, and RNA transcripts from this gene enter the egg. As these transcripts are translated, a gradient of concentration of the protein arises, with its high point anteriorly. At the same time, a second group of maternal genes, *oskar*, are active in nurse cells at the posterior pole and give rise to a second gradient with its high point posteriorly.

These two gradients activate three genes, *hb*, *Kr* and *kni*, known collectively as gap genes, in the nuclei of the developing embryo. Thus the embryo becomes differentiated into four anteroposterior regions, differing in which of the three

gap genes are active. The analogy with the flower disc of *Arabidopsis* is remarkable, although the resemblance is due to functional convergence and not to ancestral homology.

There are, however, two ways in which the primary anteroposterior differentiation of *Drosophila* may prove to be atypical:

- The early development of the *Drosophila* embryo is syncitial. The original zygotic nucleus divides repeatedly, and the resulting nuclei migrate out to form a layer lying close to the egg membrane, before any cell walls are formed. Thus the regional specification brought about by the gap genes occurs in a syncitium, and can therefore be mediated by DNA-binding proteins that do not have to cross cell walls. In contrast, in most embryos spatial differentiation requires signalling between cells, either by secreted substances or by transmembrane proteins.

- The first asymmetry in the *Drosophila* egg, between front and back, is imposed by the mother. In the frog egg, there is an early differentiation between the 'animal' and 'vegetal' pole, differing in yoke concentration and in other ways. This arises when the egg is still in the ovary, but neighbouring eggs do not have the same orientation. It seems that the initially symmetric distribution of substances in the cell is unstable, and that the asymmetry arises intrinsically, and not because it is imposed maternally. A similarly intrinsic instability causes the appearance of a dorsoventral asymmetry. It is likely that such an intrinsic origin of asymmetries is typical.

A second well-understood example of a chemical gradient that causes morphological differentiation concerns the development of the chick limb. In normal development, there appear three morphologically distinguishable digits, numbered 2, 3 and 4 to indicate their homologies to the digits of a five-fingered limb. If a small group of cells from the posterior margin of the limb bud, called the polarizing region, are removed and grafted onto the anterior margin of a second limb bud, that bud will develop six digits, which can be identified morphologically as 4, 3, 2, 2, 3 and 4. The obvious interpretation (Fig. 14.6) is that the polarizing region produces a morphogen, which, at increasing concentrations, induces digits 2, 3 and 4. This diffusing morphogen is now known, with some confidence, to be retinoic acid: a bead soaked in retinoic acid will mimic the effects of a grafted polarizing region.

14.4 Segmentation as an example of further elaboration

Figure 14.5 shows a *Drosophila* embryo just before the nuclei are separated by the formation of cell walls. In seven evenly spaced rings of nuclei, the gene *hairy*

is active, and in seven intervening rings the gene *even-skipped* is active. These 14 rings of cells form the basis for the 14 segments of the adult insect: three posterior head segments, three thoracic segments and eight abdominal segments. The process whereby these rings of cells are transformed into segments is complex. It depends on a number of 'pair-rule' genes. We will consider only the formation of the rings themselves, which seems to require three of the pair-rule genes—*hairy*, *runt* and *eve* (*even-skipped*).

In principle, there seem to be two possibilities:

Morphogenetic waves. The stripes arise from a standing wave of concentration of the kind suggested by Turing, with the peaks and troughs eliciting *hairy* and *eve* activity. A slight modification of Turing's model, involving gene

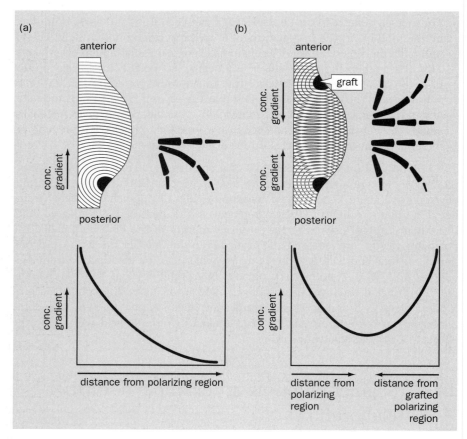

Fig. 14.6 Positional information in the development of the chick limb (after Wolpert, 1991). (a) Normal development; (b) after an additional polarizing region has been grafted anteriorly.

activation and repression, is easy to imagine. Meinhardt (1986) suggested just such a model for segment formation in *Drosophila*. And yet, as we will see, it is unlikely that the rings of Fig. 14.5 in fact arise by a periodic process.

Specific gene activation, with different signals eliciting different stripes. It is not plausible to imagine that a single gradient—for example, of the maternal *bicoid* gene product—could provide the necessary positional information. This would require that there are seven concentrations that switch on the *hairy* gene, separated by seven concentrations that switch on the *eve* gene. Although logically possible, the idea is not seriously entertained by anyone working in the field. It is possible, however, that the pair-rule genes are responding specifically to gradients set up by the gap genes, *hb*, *Kr* and *kni* (for a review, see Akam, 1989a). The transcripts of these genes are confined to specific regions of the embryo, that either abut or overlap only slightly: they could not, therefore, provide the information needed to specify 14 stripes. However, the protein products of these genes occur over a wider area, and are effective at very low concentrations: it is therefore possible that gradients of the concentrations of these three gene products, and perhaps of the products of the maternally active genes, could provide the needed information.

The reason for thinking that genes respond to specific signals is that the three relevant pair-rule genes, *hairy*, *runt* and *eve*, contain several discrete promoters, each mediating the generation of different parts of the pattern. For example, *hairy* mutants with deletions of different parts of the 5′ regulatory region cause segment fusions in different but specific parts of the embryo. There is also evidence that the *eve* gene can be switched on in more than one way; regulatory mutants are known that activate either stripe 3, or stripes 2 and 7. Finally, it is known that the gap genes encode zinc finger proteins whose binding sites are clustered in the regulatory regions of the pair-rule genes. It seems, then, that the genes *hairy*, *runt* and *eve* respond to different signals when specifying different segments; it follows that the 14 segments are qualitatively different from the beginning, and not identical, as they would be if a Turing-type process was involved.

It may be that a periodic process of the kind proposed by Turing stabilizes and amplifies the segment pattern, but there is no evidence that it is responsible for the periodicity in the first place. It will be very interesting to learn how the four petals of *Arabidopsis*, or the five of *Antirrhinum*, arise.

Segment formation, then, depends on two processes, signal emission and localized gene differentiation, which act in turn. The emission of a single signal, forming a gradient, can elicit only two, or a small number, of different states of gene activity. These states are then stably inherited, giving rise to different territories of gene expression. These in turn can give rise to new gradients, in smaller domains, which then specify another set of territories of gene expression,

and so on, until this alternation leads to the appearance of the required fine details.

14.5 From Cartesian to polar coordinates: the generation of proximodistal structures

Before data from molecular genetics were available, extensive studies were made of the regeneration of partially amputated legs of amphibians and cockroaches, and, more recently, of the wing and leg imaginal discs of *Drosophila*. These studies led to the formulation of the 'polar coordinate model' that could account for the experimental results (Figs 14.7 and 14.8). This was a formal model,

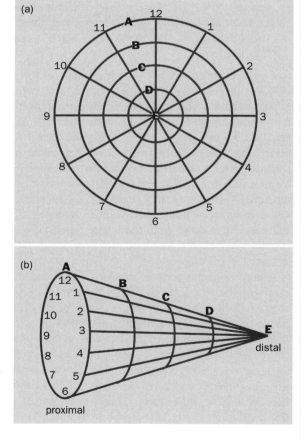

Fig. 14.7 Positional information in the polar coordinate model (after French *et al.*, 1976). (a) Cell position is determined by the radial (A–E) and circumferential (1–12) coordinate values; (b) mapping of the initial disc onto the proximodistal structure of an appendage.

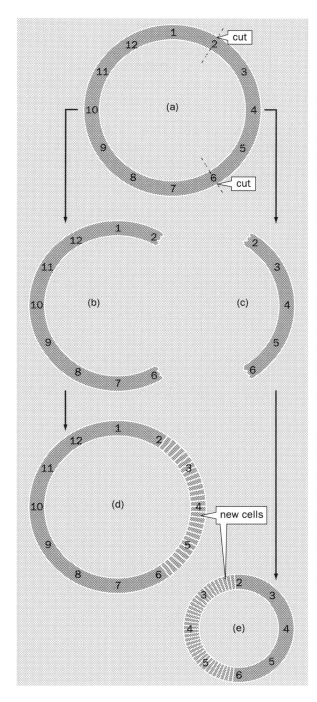

Fig. 14.8 A model of regeneration of an insect limb (Bryant *et al.*, 1981). (a) Cross-section of the limb, showing circumferential positional information. (b) Part of the limb after a lesion, with cuts at positions 2 and 6. The limb regenerates by intercalation to reform a complete ring (d). The positional values of the new cells are determined by the 'shortest intercalation rule', according to which the cells at the cut retain their values, 2 and 6, and the new cells take values connecting these by the shortest route, 3, 4 and 5 (rather than the longer alternative, 1, 12, 11, 10, 9, 8 and 7). The result is a ring of cells with the full complement of positional values, which therefore regenerates a complete limb. (c) Represents a smaller section of a limb after a lesion, but with cuts at the same positions, 2 and 6. It also regenerates by the shortest route, 3, 4 and 5, giving the structure (e), which has some values duplicated and some absent; (e) therefore regenerates as a mirror-image duplication of part of the limb.

expressed in terms of the spatial distribution of quantities whose physical or chemical nature was unknown. It consisted of a set of simple rules specifying how these quantities change when the system is disturbed experimentally. Its interest lies in two features. First, it is able to explain a surprisingly wide array of phenomena, some of which were, at first sight, decidedly odd. Second, molecular studies of the imaginal discs are beginning to provide evidence concerning the genes involved.

In the imaginal leg disc, the parts with different fates are arranged in concentric rings: the outermost ring gives rise to the coxal, and the innermost to the tarsal structures. French *et al.* (1976) suggested that the positional information is as shown in Fig. 14.7. Each cell has information about its position on the radius, A–E, and on the circumference, 1–12. When part of a disc is surgically removed, the remaining fragment either regenerates in full, or produces a symmetric, mirror-image duplication of part of the limb. Bryant *et al.* (1981) proposed an explanation for this odd phenomenon, which is explained in Fig. 14.8.

As the authors say, their model resembles Mendel's laws before the discovery of molecular genetics: it is a set of formal rules, without any molecular explanation. More recently, genetic evidence has been obtained (Bryant, 1993). Two points have been established. First, genes have been found whose spatial expression corresponds both to the circumferential and the proximodistal gradients assumed in the model. Second, some of the same regulatory genes that are involved in segment formation are also active in the limb. As yet, however, we do not know how the positional information is set up. In this respect, the study of the limb disc is in the same stage as that of flower development: we know that different regulatory genes are active in different places, but not how this localization comes about.

15 Development and Evolution

15.1 Introduction

In the nineteenth century, ideas about development, heredity and evolution were inextricably mixed up, because it seemed natural to suppose that changes that first occurred in development could become hereditary, and so could contribute to evolution. This was not only Lamarck's view but Darwin's, expressed in his theory of pangenesis. Weismann liberated us from this confusion, by arguing that information could pass from germ line to soma, but not from soma to germ line. If he was right, geneticists and evolutionary biologists could treat development as a black box: transmission genetics and evolution could be understood without first having to understand development. Since Weismann, developmental biology has had only a rather marginal impact on evolutionary biology. One day, we have promised ourselves, we will open the box, but for the time being we can get along very nicely without doing so.

Recent progress in developmental genetics, some of which has been reviewed in the last three chapters, oblige us to reopen the question. In fact, there are three related questions, not one. The first, which is most relevant to the theme of this book, is the 'levels of selection' question: why does not selection between the cells of an organism disrupt integration at the level of the organism? This is the topic of section 15.2. The second is the problem of the inheritance of acquired characters. This old problem has reappeared in a new guise. We now recognize the existence of cell heredity, mediated by different mechanisms from those concerned with transmitting information between generations. In section 15.3, we discuss whether cell heredity plays any role in evolutionary change.

Finally, in sections 15.4 and 15.5, we ask whether recent molecular information sheds any light on another old problem—that of the extraordinary conservatism of morphological form, maintained despite dramatic changes of function. This conservatism has led anatomists to identify a small number of basic archetypes, or bauplans. There is little doubt that conservatism is real. Consider, for example, the fact that bones and cartilages, which in humans serve in swallowing, sound production and hearing, are derived from elements of the gill apparatus whereby our fish ancestors exchanged gases with seawater, and, before that, in all probability, from elements of a filter-feeding apparatus. The discovery of such morphological homologies is a triumph of comparative anatomy, but their meaning is less clear.

The simple explanation is that bauplans were once functional adaptations to particular ways of life. For example, the segmented musculature, notochord, paired fins and post-anal tail that are fundamental to the chordate bauplan evolved in the first place as adaptations for sinusoidal swimming. Later ecological adaptations—for example, walking and flying—were achieved by modifying existing structures rather that by creating new ones. Given the stepwise nature of evolution, that is not surprising. Yet, since the time of Goethe and Saint-Hilaire, some anatomists have felt that bauplans represent something more than

frozen historical accidents: in some way, they reveal fundamental laws of form. Molecular data have introduced a new twist into this old debate. It now seems that the genetic signalling system of animals may be more ancient, and more conservative, than the morphology that it determines.

15.2 Development and the levels of selection

How far has a conflict between selection at the between-cell and the between-organism levels influenced the pattern of development? Buss (1987) argued that the influence has been pervasive. His general approach to the major events in evolution is similar to our own. He suggested that there have been a series of major transitions between different units of selection, and gave a list of these transitions that is almost identical to ours. The first outline sketch of our book (Maynard Smith, 1988a) was much influenced by his thesis. It is of particular importance that we should acknowledge our debt to Buss because, when it comes to the details, we are unable to agree with him.

Buss was primarily concerned with the transition from single-celled to multicellular organisms. We argued in section 1.4 that the crucial reason why competition between cells does not disrupt the organism is that, typically, development starts from a single cell, so that, apart from somatic mutation, the cells of an individual are genetically identical. Buss did not mention this fact. Instead, he emphasized two aspects of development—maternal control and the early segregation of the germ line—that he saw as preventing, or making less likely, the spread of selfish mutations. The argument is as follows. Imagine a selfish mutant, S, occurring in a somatic cell, and causing that cell to be more likely to give rise to a germ cell, but at the expense of lowering the fitness of the organism. Genes in the mother of the individual in which the mutation S occurs, or at other loci in the individual itself, increase their chance of future survival if they can suppress the effects of S. Genes in the mother can do this by extending maternal control of early development. Genes in the individual can do so by limiting the capacity to become a germ cell to a specific cell lineage: that is, by early segregation of the germ line.

The two mechanisms proposed need separate discussion. Consider first maternal control. If the whole development of the offspring was controlled by the mother, then no S mutant could be effective in the offspring: indeed, no genes in the offspring would affect its phenotype. We would be faced with a curious system in which the phenotype of each generation was determined by genes in the previous one. It would be an interesting exercise to work out the consequences of such a system. In practice, however, the extent of maternal control is very limited. Even in *Drosophila*, in which the extent is perhaps greater than in any other species that has been adequately studied, it could have only a very limited effect in reducing the damage caused by S mutations in the offspring. The

significance of maternal control in flies, as well as of some other curious features of their development, such as the syncitial nature of early cleavage, probably lies in the fact that many of them are adapted to life in evanescent habitats—for example, rotting meat, fruit and seaweed, and temporary water bodies. Maternal control may help to ensure rapid development.

Turning to the significance of the early segregation of the germ line, it is useful to distinguish two kinds of somatic mutation: *S* mutants, that increase the chance that a cell will give rise to a gamete, and *M* (malignant) mutants, that cause unregulated cell proliferation, with a consequent lowering of individual fitness. There will be no selection on genes at other loci to suppress *S* mutants *per se*, unless they are also malignant, or in some other way lower individual fitness. The main selective force, therefore, will be to prevent or delay malignancy. This would not be helped by segregation of the germ line. It is, however, possible to imagine *S* mutants that are not malignant, but that lower fitness. Imagine, for example, a somatic mutation in a plant meristematic cell, making the cell more likely to give rise to a flower rather than a leaf. In general, such mutations would disturb the optimal balance between leaf and flower production. Doubtless they occur, but they are prevented from becoming common by individual selection. Individuals arising from the gametes carrying the new mutation will have too few leaves and too many flowers. But note that such selection works only if plants develop from a single cell. There is, of course, no germ-line segregation in plants.

Early segregation of the germ line is precluded in plants, because plant cells do not move, and so there is no way in which cells from a segregated lineage could be transported to the ovules or pollen. But why is it so common in animals? Two suggestions can be made, although there is little direct evidence for either:

- There has been selection to reduce the mutation rate in gametes. This is most easily achieved if gametes are produced by a specialized lineage with no other functions.

- When gametes are produced, all genes must be reset to a totipotent state. There will be fewer errors in this process if fewer genes undergo differentiation in the first place.

Either, or both, of these selective advantages could have lead to early germ-line segregation. There is, therefore, no reason to think that either maternal control or germ-line segregation evolved to suppress cell–cell competition. But the question is important, and we do not regard our arguments as decisive.

We find it even harder to accept Buss's second thesis in regard to these processes. He argued that, as the window between maternal control and germ-line segregation narrowed, the evolution of radically new body plans became difficult or impossible. His reasoning was that important evolutionary novelty requires that mutations be synergistic, in the sense of being selected both at the between-

Fig. 15.1 A model of 'synergistic selection': germ-line cells are shaded; somatic cells are unshaded. Two evolutionary states are considered: in A, only two cells in the blastula contribute to the germ line; in B, four blastula cells contribute. Consider, first, organism-level selection favouring evolution from A to B. In state A, cells have the property X: 'if in position 1, become a germ cell'; what is required is a mutation to property Y: 'if in position 1 or 2, become a germ cell'. Selection favours Y synergistically: a new mutant is more likely to enter a germ cell (if it occurs in cells U), and zygotes carrying Y are fitter. But the mutation Y would spread by between-individual selection, in the absence of within-individual selection (e.g. if the mutation occurred only after gastrulation). Thus the effect of synergistic selection is only to increase the effective mutation rate (i.e. the mutation rate in gametes). Now consider selection favouring the reverse change, B to A. What is required is a mutation Y to X. A new mutant will have an opportunity of spreading only if it occurs in cells V, or their descendents. The effect of contrary within-individual selection is, therefore, only to reduce the effective mutation rate: it will not prevent evolution from B to A.

cell and the between-organism levels. We cannot see why this should be so. Consider a mutation occurring in a germ-line cell, which has no effect on that cell or its immediate descendents, but which influences morphogenesis if it is present in somatic cells in the next generation. Such a mutation can spread by natural selection. Germ-line segregation is irrelevant. The question is analysed

further in Fig. 15.1. Much has been made of the fact that few, if any, wholly new body plans have emerged since the Cambrian, but it is hard to see how it could be otherwise. Organisms with an already evolved body plan could hardly give rise to descendents with a completely different one, so the only possibility is that new phyla should have evolved from single-celled ancestors. The prior existence of many highly elaborated multicellular animals has apparently prevented this.

15.3 Cell heredity and the inheritance of acquired characters

Given the existence of cell heredity, it is natural to ask whether states of gene regulation can occasionally be transmitted through the sexual process, and, if so, what significance this has for evolution. Holliday (1987) was one of the first to raise these questions, which have been discussed more recently by Jablonka & Lamb (1989) and Jablonka *et al.* (1992). The clearest evidence that epigenetic characters can sometimes by transmitted through the germ line is afforded by the phenomenon of genomic imprinting. In mammals, for example, homologous chromosomes or genes may be differently methylated, depending on their parental ancestry. At some loci, only the gene inherited from the father is active, and in other tissues only the gene inherited from the mother. This implies the existence of different labels on paternally and maternally inherited genes. Since assortment at meiosis is random, this means that maternal chromosomes in the germ line of a male must be relabelled, as must paternal chromosomes in a female. The germ line is therefore accessible to re-programming, and the new labels can then be transmitted through meiosis.

Other examples of the inheritance of epigenetic modifications include the effects of fertilizer treatment in flax (*Linum*), and the alternation of heteromorphic generations in invertebrates. It has been known for some time (Durrant, 1962) that flax plants treated with fertilizer have a more branched pattern of growth and broader leaves. These traits are seed-transmissible: it turns out that the cells of treated plants respond by amplification of specific regions of the genome (Cullis, 1983). Turning to the alternation of generations, in digenetic flukes, for example, a single cell, with an unaltered diploid genome, can develop into a miracidium, a redia or a cercaria, depending on its recent epigenetic history: of course, in this case the epigenetic state is not transmitted through meiosis, but it does pass through a single reproductive cell.

Given that, at least occasionally, epigenetic states are sexually transmitted, what are the evolutionary consequences? In an attempt to answer this question, Maynard Smith (1990) developed a formal model of what he called a 'dual

inheritance system'. The model made the following assumptions:

- All genes, structural and regulatory, have specific sequences determining their activity. A sequence may be labelled, in which case the gene is inactive; unlabelled genes are active. Labels are copied when cells divide.

- The presence or absence of a label on a specific sequence can be altered by the activity of a labelling gene, which codes for a protein that can recognize the specific sequence.

- In turn, the activity of these labelling genes is determined by a set of regulatory genes: the protein produced by such a gene is effective in switching on a labelling gene only if the protein is bound to an appropriate inducer. An inducer can therefore alter the state of a cell by combining with a regulatory protein, which in turn activates a labelling gene whose product, in its turn, alters the labels on many genes in the cell, and hence their activity.

The model sounds, and is, rather complicated, but it is perhaps as simple as it can be, and still explain the phenomena of differentiation and cell heredity. Also, it does not assume anything for which there is no experimental evidence.

Given such a model, it is rather easy to design systems with the required behaviours: details are given in the original paper. In particular, the following systems are possible:

- Germ cells derivable from somatic cells, as in plants and fungi (Fig. 15.2). In the absence of external induction, germ cells give rise to germ cells, and somatic cells to somatic cells. Inducer I_2 converts somatic cells to germinal cells, and I_1 converts germinal cells to somatic cells.

- A system resembling (1), but with two types of somatic cell, A and B, induced by I_A and I_B, respectively (Fig. 15.3a). Once differentiated, somatic cells of either type cannot be converted to the other, but each can be converted to germinal cells by I_1. States A and B can be thought of as different tissues, or as different adults that have adapted to different environments during development.

- A system reminiscent of that in *Linum* (Fig. 15.3b). As in Fig. 15.3a, germinal cells of type 1 (GL1) can give rise to either type of soma, A or B, depending on

Fig. 15.2 The behaviour of a hypothetical control system (from Maynard Smith, 1990). The system resembles that in plants: germ-line cells can be converted to somatic cells by inducer I_2, and somatic cells into germ cells by inducer I_1.

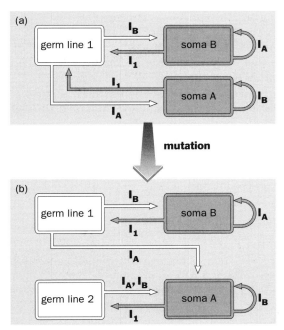

Fig. 15.3 The behaviour of a more complex control system (from Maynard Smith, 1990). System (a) represents a higher plant, with two types of differentiated somatic cell, each able to give rise to germ cells. A single mutation can convert this into (b), which resembles the situation described in flax (*Linum*). I_A and I_B represent different environments, and soma A and soma B different adult phenotypes. Soma B breeds true in environment $I_B'^\wedge$ but is converted to soma A if development is in environment I_A; soma A breeds true in either environment. Thus system (b) illustrates the 'inheritance of an acquired character', not involving any change in DNA sequence.

induction. But although soma B can be converted to GL1, soma A can give rise only to a different type of germ line, GL2, which gives rise to soma A regardless of the type of inducer. Thus, starting with GL1, the inducer I_A produces a changed phenotype that is subsequently inherited indefinitely—the inheritance of an acquired character.

Evolution from the system in Fig. 15.3a to that in Fig. 15.3b requires only a single mutation altering the specificity of one of the labelling genes. What is the evolutionary relevance of all this? Suppose that, in Fig. 15.3, I_A and I_B represent different environments, and soma A and soma B are adults adapted to those environments. Then, if it became adaptive to develop soma type A regardless of the environment, without waiting for induction, natural selection would favour any mutation converting 15.3a to 15.3b. An adaptation that originally occurred during development would have become genetic: this is precisely what Waddington (1956) had in mind when he coined the term 'genetic assimilation'.

Such a process is interesting, but we do not think it alters greatly our view of evolution. Natural selection, acting on mutations that are more likely to be non-adaptive than adaptive, is still the fundamental process. In the initial state (Fig. 15.3a), therefore, it is only because of past natural selection that development is adaptive: it would be equally possible to have DNA sequences, and the resulting control processes, that caused soma A to develop in response to I_B, and soma B in response to I_A—that is, a maladaptive response. It is worth remembering that

when Waddington demonstrated genetic assimilation experimentally, he studied the assimilation of a maladaptive response (absence of a cross-vein induced by heat shock). Further, the adaptive evolution from 15.3a to 15.3b occurs by the selection of the precise mutation that brings about the desired transition: there are many possible mutations that do not produce an adaptive response.

Jablonka *et al.* (1992) questioned the validity of this argument. They proposed a mechanism whereby epigenetic changes could be stably inherited through sexual reproduction without the need for changes in DNA sequence. However, their model requires that there be alternative, replicable marking patterns irrespective of the genetic content of the genes they mark. Such marking patterns would be autocatalytic, and have their own ontogenesis. We do not know how frequent such situations are in real organisms.

15.4 Gene homology in development

The 'homeotic' mutants of *Drosophila*, such as *antennapedia*, which causes leg-like appendages to grow where there should be antennae, or *tetraptera*, which causes the halteres to be replaced by a second pair of wings, have been known for 50 years. Their characteristic feature is that they cause the appendage appropriate to one segment to appear in another. Recent molecular analysis (reviewed by Akam, 1989b) has revealed some fascinating, and at first sight unexpected, facts. The homeotic genes in *Drosophila* occur in two tightly linked clusters, together constituting the 'homeotic gene complex', or *HOM-C*. The split into two unlinked clusters is probably recent, since in the beetle, *Tribolium*, the homologous genes occur in a single cluster.

All homeotic genes contain a DNA-binding region called the homeobox. It consists of 60 amino acids, forming a helix-turn-helix structure resembling that of bacterial repressor proteins. The genes are active in specific segments, and are thought to be responsible for the appropriate development of those segments. The sequence of their activity along the anteroposterior axis of the embryo is identical to their sequence along the chromosome.

A homologous set of genes, the *Hox* genes, has been found in the mouse. There are four linked clusters of *Hox* genes: two of these are complete duplications of the whole set, the other two being incomplete. The existence of more than one cluster may indicate that these genes are concerned in more than one developmental process. When the sequences of the homeobox domains are compared, it is possible to find *Hox* genes homologous to particular *HOM-C* genes. It turns out that the *Hox* genes occur in the same order along the chromosome: the order of their activities along the anteroposterior axis is also the same (Fig. 15.4).

It has since been found that homeobox domains occur in genes that regulate development, and that similar domains are present in the genes that determine

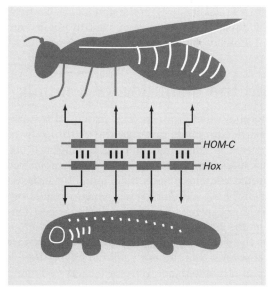

Fig. 15.4 Relation between the *HOM-C* genes in *Drosophila* and the *Hox* genes in the mouse. The vertical lines indicate the closest homologies of the homeobox domains; the arrows indicate the regions of the embryo in which the genes are active: the anteroposterior order is the same as the order on the chromosome. The figure shows relationships, not the precise locations of gene action.

yeast mating type. The latter is not all that surprising, since the mating-type genes in turn specify other gene activities concerned with pheromone production and recognition, and with cell recognition. We conclude that homeobox domains are concerned with switching on cascades of other genes, and have been so since early in the evolution of eukaryotes.

But why should there be a precise correspondence between position along the chromosome and site of action along the anteroposterior axis? Even if this correspondence existed in the common ancestor—and it is not obvious why it should have done—why should the gene order have been retained? One possibility is that there is some *cis*-acting control passing along the chromosome between neighbouring genes. Two observations argue against this. It has already been mentioned that, in dipterans, the genes have been divided into two clusters, which would be unlikely if *cis*-acting signals were essential. More directly, transgenic mice have been constructed with homeotic genes in unusual chromosomal locations. By using reporter genes controlled by the same promotor, it has been found that these genes are active in the appropriate cells, and not elsewhere. Despite these observations, however, it seems likely that there is some functional sense in the conserved chromosome order.

The power of gene expression in solving old questions in comparative morphology is illustrated by a report that appeared while this book was in proof (Arendt & Nübler-Jung, 1994). It turns out that there is homology between genes determining dorsal structures in annelids and arthropods, and genes determining ventral structures in vertebrates. This revives an old, and much de-

rided, theory first proposed by Geoffroy St Hilaire early in the nineteenth century. This is that vertebrates, with their dorsal nerve cords and ventral hearts, are indeed upside-down descendents of annelid-like ancestors.

15.5 The zootype and the definition of animals

Animals are usually defined on morphological grounds. The structural features common to the Metazoa, or to particular phyla such as the Arthropoda or Vertebrata, have even been interpreted as revealing some fundamental, if unknown, 'laws of form'. Recent work in developmental genetics, briefly described in the last section, is leading to a rather different picture. In all animals studied so far, from cnidarians to higher metazoans, a homologous genetic regulatory system for relative positional information seems to be present. This is best illustrated by the *Hox* gene cluster, or the homologous *HOM-C* cluster, although other examples of conservation in the pattern of expression of position-coding genes are also known. The essential point is that, in many animal phyla, there is a stage in development when a series of homologous genes are expressed along the anteroposterior axis, specifying the morphological structures that will later develop. Although different structures in fact develop in different phyla, the signals that elicit them are similar.

In all probability, this system for relative position is very ancient, and was present in the the common ancestor of all animals. This has led Slack *et al.* (1993) to propose that this character is the defining synapomorphy (shared ancestral trait) of the kingdom Animalia, and should be called the zootype (Fig. 15.5).

The discovery of a signalling system common to all animals is recent. It must be related in some way to a feature of development that has been recognized for some time: this is that, during development, the members of a phylum often pass through a common morphological stage, even though the processes that give rise to this stage differ between members of the phylum. An example is the planula larva of cnidarians. Despite very different patterns of cleavage and gastrulation, the members of the phylum converge on a common form, before diverging again. In vertebrates, there is a similar convergence on the pharyngula stage.

In some cases, it is easy to see why the early stages differ: for example, the pattern of cleavage may depend on the amount of yolk in the egg. What is less obvious is why the various types should then converge morphologically. A second puzzling feature is that the form upon which they converge may well resemble that of an adult ancestor. Thus the planula larva is often a free-living stage that may resemble an ancestral cnidarian, and many anatomists think that the ancestral chordates were sinusoidally swimming filter-feeders, with an anatomy resembling that of the pharyngula stage of modern vertebrates.

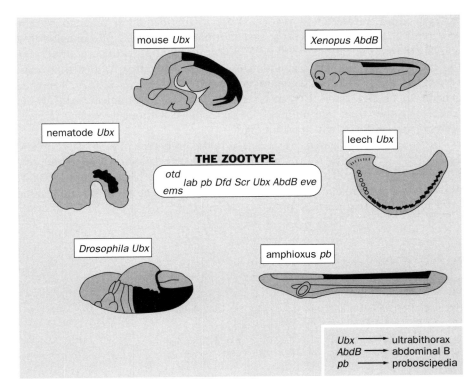

Fig. 15.5 The zootype (after Slack *et al.*, 1993). The diagram in the centre shows the spatial pattern of activity of genes in the *Hox* gene cluster, and some other genes, using the *Drosophila* nomenclature. Homologous genes are now known to be expressed in a similar pattern in embryos from several phyla. The drawings of the mouse, *Drosophila*, nematode and leech embryos show the pattern of activity of the *Ubx* gene and its homologues, as seen in whole mounts: the drawings of *Xenopus* and amphioxus show the activity of other genes in the series.

Slack *et al.* (1993) suggest the term phylotypic stage for this common embryonic form; in fact, they suggest that it can be characterized in three ways:

- The stage at which all major body parts are represented in their final positions as undifferentiated cell groups.
- The stage after the major morphogenetic cell movements.
- The stage at which all members of the same phylum show the maximum degree of resemblance.

They also suggest that the phylotype is the stage at which the zootype—that is, the spatial expression of regulatory genes—is most apparent. Although this seems to be true, the reasons are not obvious, at least to us.

There are some profound consequences of this definition of animals. The zootype is based on functions that are informational: it is easy to imagine that different systems could have evolved in different phyla. Thus the organization of the zootype is gratuitous, and indicates common ancestry, rather than convergent adaptation or developmental constraints. Although arbitrary, it may be difficult to change once it is there: other parts of development, concerning processes rather than information, can be modified without disrupting the underlying system for relative positioning.

This concept of the zootype revives that of the archetype of animals, proposed by Geoffroy de St Hilaire. But this archetype does not seem to be a uniquely stable state of matter. This observation is a rather strong blow at the essentialist or structuralist approach to body plans: the definition of animals rests on genetic ancestry, rather than on first principles of form.

16 The Origins of Societies

16.1 Introduction

The heat generated within a mound of the termite *Macrotermes* is carried upwards by a central air duct (Fig. 16.1). The air then travels down along narrow channels close to the surface of the mound, where it is cooled, and where, as in a lung, oxygen and carbon dioxide are exchanged. The whole mound is an air-conditioning system. Although the mound resembles a human building in having features ensuring the comfort of its inhabitants, it differs in that not one of its builders had a picture of the completed structure before building started. The structure emerged from the rule-governed behaviour of tens of thousands of interacting workers. In this, the mound resembles a human body rather than a human building. The body is built by the rule-governed actions of millions of cells. Nowhere is there anything resembling a blueprint of the body. At most, the genome is a set of instructions for making a body: it is not a description of a body.

The resemblance between the development of an insect colony and of an organism has led to the concept of a 'superorganism'. The analogy has some value. To the extent that individual ants, bees or termites have lost the capacity to reproduce, they can propagate their genes only by ensuring the success of the colony, just as somatic cells can propagate theirs only by ensuring the success of the organism. Hence, the colony can be expected to have features adapted to ensure its success, and it is reasonable to apply concepts of optimization to it,

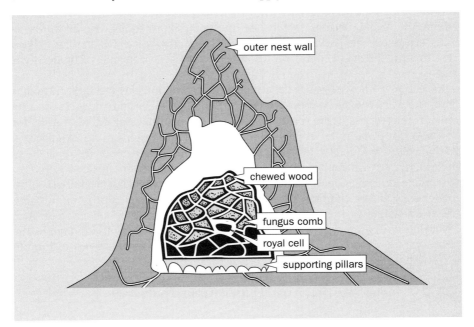

Fig. 16.1 Section of a termite mound (after Grassé & Noirot, 1958).

rather than to the individual—as was done, for example, by Oster & Wilson (1978) in their book on insect caste systems.

But for our purposes the concept of a superorganism is of little use. To understand the origins of animal societies, we must ask how individuals capable of reproduction came to cooperate to the extent that most of them lost the ability to reproduce. To understand their maintenance, we must explain why they are not disrupted by cheating. Unlike somatic cells, the individual workers, although related, are not genetically identical. We would therefore expect within-colony conflict to be widespread, as indeed it is: examples discussed below concern egg-laying by workers, and conflict over the sex ratio. As with the other transitions we have discussed, the problem is to explain how this conflict is kept within bounds.

In section 16.2, we discuss three main processes—kin-selected altruism, enforcement and mutual benefit—that favour cooperation. In section 16.3, we describe briefly the kinds of animal society, and their taxonomic distribution, and speculate about the routes that may have been followed to sociality. In section 16.4, we discuss a series of genetic topics. These are only rather tenuously related to one another. We first ask how egg-laying by insect workers is regulated. This is best seen as an illustration of how kin selection and enforcement operate to limit conflict. We then turn to the role of haplodiploidy in the origin of social behaviour in the Hymenoptera, and conclude that there may well be a causal connection between them. This raises the question of why termites, which are diploid, became social. The question is hard to answer, because termites are all eusocial, and have no primitively social relatives which might give a clue to their history. Finally, we ask whether social insects can recognize their kin.

In section 16.5, we discuss the ecological preconditions for sociality. Although brief, the point made is a crucial one. The evolution of sociality does not depend only on relatedness: there must be something useful that individuals can do for one another. Finally, in section 16.6, we discuss the origins of human society. As biologists, we do not aim to write a book about anthropology or political philosophy, but we cannot simply omit the last of the major transitions. We try to confine ourselves to the following question: what biological changes have made it possible for humans to construct types of society different from anything found elsewhere in the animal kingdom? We think that the crucial biological invention was language. We leave discussion of the nature of language to the next chapter, but, here, we touch on its significance for the types of human society that can develop.

16.2 The evolution of cooperation

We now consider in more detail the three processes—kin-selected altruism, enforcement and mutual benefit—that can lead to the evolution of cooperation.

They are not alternatives: to varying degrees, all are important in all societies, including human ones.

Kin-selected altruism

The basic idea is that, by helping a relative, an individual is propagating its own genes, or, more precisely, copies of those genes. It was hinted at by Darwin, and expressed in genetic terms by Haldane (1955). However, its precise formulation, and its application to animal societies, we owe to Hamilton (1964 and subsequently). It is most easily understood in terms of 'Hamilton's inequality', which states that a gene causing an individual to perform an act will increase in frequency in a population if $b/c \equiv k > 1/r$, where b is the benefit to the recipient and c the cost to the actor, both measured as changes in the expected number of offspring resulting from the act, and r is the relatedness of the actor to the recipient (Box 16A).

Although apparently simple, Hamilton's idea has been widely misunderstood. Two misunderstandings are sufficiently widespread to be worth mentioning. The

BOX 16A Hamilton's inequality

Imagine a rare gene, A, which causes an individual D (donor) to perform an act X. (This formulation has been criticized on the grounds of 'genetic determinism': however, the argument works just as well if gene A makes it more likely that, in particular circumstances, the individual will do X.) The effects of act X are:

- to reduce the expected number of gametes that D passes to the next generation by c (the 'cost'); and

- to increase the expected number of gametes that a 'recipient', R, passes to the next generation by b (the 'benefit').

In a diploid organism, the effect of the act is to reduce the number of A genes in the next generation by $c/2$ (because each gamete has an evens chance of carrying A), and to increase the number by pb, where p is the probability that a gamete produced by R will carry gene A. If R is unrelated to D, then p is zero, because we have assumed that the gene is rare. The 'relatedness', r, of individual R to D is defined as the proportion of genes in R that are 'identical by descent' to genes present in D: 'identical by descent' means inherited from a recent common ancestor. If D is a diploid, $p = r/2$, because a random gene at the locus in R has a chance r of being present in D, and hence a chance $r/2$ of being identical by descent to A.

The act therefore increases the number of A genes in the next generation if $rb/2 > c/2$, or $b/c > 1/r$, which is Hamilton's inequality. The argument above applies only to a rare gene: Hamilton (1964) showed that it holds equally well for a gene that is common in the population. Grafen (1985) gives a lucid geometrical explanation.

first is conveniently referred to as Sahlins' fallacy (1976). It is that, for kin selection to operate, the actor must be able to calculate its relatedness to the recipient. In fact, if a gene is to cause the spread of a gene in context X (e.g. between nest-mates), all that is needed is that the inequality should hold for the average value of *r* in that context (0.5 if nest-mates are always full sibs, and somewhat less if there is egg-dumping or adultery): no-one has to know what *r* is, except the biologist trying to test Hamilton's idea. The second fallacy is more excusable. Examples of it are that humans should be particularly nice to chimpanzees because they share 98 per cent of their genes, and that one female of the asexual species of lizard, *Cnemidophorus uniparens*, should be particularly nice to others, because they are genetically identical (in fact, they attack and kill one another). To see why this is wrong, consider the spread of a new mutation, *A*, causing female lizards to act altruistically. We are concerned only with the spread of gene *A*, relative to its allele *a*: the rest of the genome is irrelevant. If gene *A* causes a female to help the survival of a random member of the population, this has no average effect on the frequency of *A*. But if the gene causes the female to help its own offspring, which are certain to have *A*, this causes an increase in *A*'s frequency.

In section 16.3, we show how the concept of relatedness can explain some surprising details of animal societies.

Enforcement

If one individual forces another to cooperate, the latter can hardly be said to have acted altruistically. Alexander (1974) argued that much social behaviour is to be explained because parents force their offspring to cooperate. Enforcement can also occur between sibs. Dominance relationships are certainly important in social insects, as the following quotation (Packer, 1986) shows: 'Large individuals were observed to take hold of smaller ones by the neck with their mandibles. The small females were then taken to the bottom of the burrow and repeatedly pummelled into the earth at the end or bashed from side to side against the burrow walls.'

Dominance and relatedness both play a role in a model of the origin of insect sociality suggested by Stubblefield & Charnov (1986). Imagine a female insect whose mother mated only once. She can choose between caring for her own offspring or her mother's (i.e. her full sibs). In both cases, her relatedness is one half, so she has no preference. But her mother is more closely related to her own offspring (0.5) than to her daughter's (0.25). It should therefore be easy for the mother to force her daughter to look after her sibs: the mother has much to gain, and the daughter nothing to lose. If the mother has several daughters, the tendency is still stronger, because each daughter will prefer her sisters to care for further siblings than for nieces and nephews.

This argument seems almost too successful. The problem becomes, not to explain sociality, but to explain why all animals are not social. One crucial

assumption we have made is the following: the daughter would be equally successful raising her own offspring or her sibs. This need not be true. It might be easier to raise her siblings (e.g. because she would not have to make a new nest) or her own offspring (e.g. because, by dispersing, she would find new resources). Whether Hamilton's inequality holds depends not only on relatedness but also on costs and benefits. This leads us to the third factor determining the evolution of sociality.

Mutual benefit

If an individual can produce, say, two offspring on its own, but a group of n individuals can raise $3n$ offspring if they cooperate, then it pays each of them to cooperate. Such synergistic effects are often invoked to explain cooperation. For example, perhaps lions are social because group hunting is more efficient than hunting alone. In fact, if success is measured per individual, as it should be, lions seem to hunt better on their own. During a group hunt, individuals tend to hang back and leave the dangerous work to others (for a review of cooperative hunting, see Packer & Ruttan, 1988). Nevertheless, synergistic effects probably are responsible for lions being social cats. A group of females can defend a kill against hyenas and other lions, whereas a single female cannot, so it pays females to live in groups. A group of males can successfully defend access to a group of females whereas single males cannot, so the breeding success of an individual male increases with the size of the group to which he belongs, even though he must share matings (Packer *et al.*, 1985).

The snag, of course, is that it takes two to cooperate. It is important to distinguish two situations, which we will call the rowing and skulling games (Fig. 16.2). The skulling game is identical to the familiar Prisoner's Dilemma. If both players cooperate, they are better off than if both defect, but, unfortunately, whatever the other player does it pays to defect, so 'rational' players end up defecting. Various attempts have been made to explain the evolution of co-operation in such games. If the game is played once against a random opponent, evolution leads inevitably to defection. But if the game is played repeatedly against the same opponent (Axelrod, 1984; for the theory of the repeated Prisoner's Dilemma game, see Box 16B), or against neighbours in a sessile population (Axelrod & Hamilton, 1981; May & Nowak, 1992), more cooperative behaviour can evolve.

However, the intellectual fascination of the Prisoner's Dilemma game may have led us to overestimate its evolutionary importance. There is some evidence that, in an experimental situation, animals do not learn to cooperate in the repeated Prisoner's Dilemma game. Rowing may be a better model than skulling of the situations in which cooperation evolves. Once it is common, cooperation is evolutionarily stable. The problem is how it becomes common in the first place, because defection is also stable. A possible answer is that cooperation starts

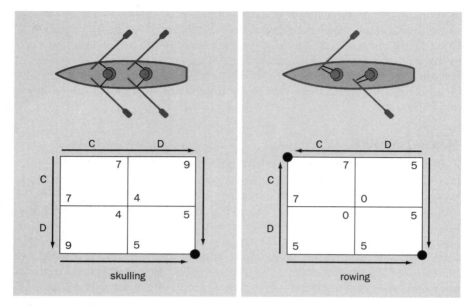

Fig. 16.2 The skulling and rowing games. It is assumed that the pay-off for travelling forwards slowly is + 4, for travelling fast is + 7, and for not exhausting oneself by rowing is + 5. The possible strategies are to cooperate, C, and to defect (i.e. not row), D. Each square in the matrices contains (bottom left) the pay-off to the strategy on the left, and (top right) to the player above. The solid circle indicates the evolutionarily stable states (ESS) of the system. The skulling game corresponds to the Prisoner's Dilemma: D is the only ESS. In the rowing game, cooperate is also an ESS.

Table 16.1 The repeated Prisoner's Dilemma game

		D	C	
(a)	D	2	4	
	C	1	3	

		D	C	TFT
(b)	D	20	40	22
	C	10	30	30
	TFT	19	30	30

between relatives, and later, once it is common, spreads to non-relatives. Returning to lions, it is interesting that, although the groups of males that hold a group of females are often relatives, it is not uncommon for unrelated males to come together for this purpose. In any case, the important point is that co-operation must sometimes pay: we return to this point in section 16.3.

BOX 16B The repeated Prisoner's Dilemma game

Two individuals play the game illustrated in Table 16.1a against one another 10 times. An individual can adopt one of three strategies:

- C: always cooperate;

- D: always defect; or

- TFT: 'tit-for tat'; cooperate in the first game, and in each subsequent game play what the opponent played in the last game.

It is then easy to calculate the total pay-offs over 10 games, shown in Table 16.1.

For the single game, Defect is the only evolutionarily stable strategy (ESS): a population playing D cannot be invaded by a C mutant. For the repeated game, D is still an ESS, but TFT is also an ESS, and has, in fact, a larger basin of attraction. TFT is a relatively cooperative strategy, because two TFT players always cooperate with one another. Thus, cooperation is more likely to evolve if an individual interacts repeatedly with the same partner.

One snag with the game just described is that, if there are always exactly 10 games, a player would be tempted to defect in the last game. This difficulty disappears if, instead of a fixed number of games, we make the more realistic assumption that, after each game, there is a fixed probability of a further game. Then a player never knows he or she is playing the last game, and must calculate according to the expected number of further games.

The problem is much more complicated if allowance is made for other possible strategies (Nowak & Sigmund, 1993).

16.3 Kinds of animal society

Societies such as those of lions, chimpanzees or baboons, in which all adults are potential breeders, require little explanation other than immediate mutual benefit: they are rowing games. One additional point is worth making because of its relevance to the origin of human society. Among mammals, all social groups are either matrilineal (males leave the natal group before breeding), or, less commonly, patrilineal. The probable explanation is that this behaviour evolved because it prevents inbreeding.

In some birds and mammals (particularly canids), the social group consists of a single breeding pair, together with other adults that help to raise the juveniles. These 'helpers' are usually offspring of the breeding pair, so they are raising their own sibs. However, helpers may later become dominant breeders, and to be a helper may increase their chance of doing so.

The most complex animal societies, however, are those in which some individuals are permanent non-breeders. Groups are defined as 'eusocial' if they contain sterile workers that help their parents. Predominantly such societies are found among insects, but the naked mole rat and several spiders have evolved a similar system. In this and the next section, discussing social insects, we have relied heavily on the account given by Seger (1991). Among insects, eusociality is found in termites and hymenopterans. The termites, which are relatives of cockroaches, are all eusocial, and so probably represent a single origin. There are many solitary hymenopterans, and sociality has evolved many times. All ants are eusocial; sociality has evolved several times among wasps, and many times in the bees.

Since we are concerned with origins, we are particularly interested in those species, such as the bee *Halictus rubicundus*, in which both social and solitary reproduction occurs. The related *H. ligatus* has a range extending from Canada to the neotropics. In the north the species is eusocial, but in the south there are communities of independently reproducing females, together with females that are brood parasites, laying their eggs in the nests of other females. A study of such primitively social bees suggests that there are two evolutionary paths to eusociality. The subsocial route starts from the 'helpers at the nest' system described above, and evolves to eusociality if the helpers become permanently sterile. The parasocial route starts with a group of fertile females cooperating to feed and defend their young, and evolves via a state in which some of these females become sterile workers to the fully eusocial condition in which many of the offspring are also workers.

The economics of the parasocial route has been investigated further. The logic has already been described: if one female produces two offspring, and n females produce $3n$ offspring, it pays to cooperate. This can account for a community of breeders. But suppose that, once the nest is established, a struggle for dominance leads to one of the n cooperating females becoming the queen, and the remaining $n - 1$ becoming workers. It would still pay to cooperate, because it is better to have a $1/n$ chance of having $3n$ offspring than to be sure of having two offspring. This arithmetic was checked by Bartz & Hölldobler (1982) in the ant *Myrmecocystus mimicus*. They found that, in the laboratory, the largest number of offspring per foundress was produced when there were three foundresses: in the wild, most colonies have two to four foundresses.

16.4 The genetics of insect sociality

Why don't worker bees lay more eggs?

In many ants, bees and wasps, workers retain functional ovaries, and are therefore able to lay male eggs: being unmated, they cannot have daughters. Yet it is rare for workers to do so, except in colonies that have lost the queen, or

never had one. This is puzzling, because a female is more closely related to her own sons than she is to eggs laid by the queen. Pheromones seem to be the immediate cause, but are not a satisfactory evolutionary explanation: why should workers pay any attention to a mere signal?

In the honey-bee, *Apis mellifera*, some workers have functional ovaries, but electrophoretic studies show that only about 0.1 per cent of drones develop from worker-laid eggs. Ratnieks (1988) has suggested a possible explanation. Females mate with many males before founding a colony. It follows (Fig. 16.3a) that a worker is more closely related to eggs laid by the queen than to eggs laid by other workers. It would therefore pay workers to destroy eggs laid by others. If so, the effective sterility of workers is caused by policing by other workers—a form of enforcement—and not only by a response to a pheromonal signal. This explanation requires that workers are able to distinguish between queen-laid and worker-laid eggs. Ratnieks & Visscher (1989) found that workers preferentially destroyed worker-laid eggs. A further prediction is that, in species in which the queen mates only once, and in which (Fig. 16.3b) workers are therefore more closely related to their sister's sons (3/8) than to their mother's sons (1/4), worker egg-laying should be more frequent. What data there are tend to confirm this.

Does haplodiploidy predispose hymenopterans to sociality?

Hymenopterans are certainly predisposed to sociality. They are also haplodiploid. Is there a causal connection? Readers are warned that this apparently simple question is among the most difficult in population genetics. A simple but probably fallacious argument goes as follows. If females mate only once, then a female is more closely related to her sisters (3/4) than to her daughters (1/2): therefore, it should pay her to stay at home and look after her sisters. The snag is that she is more distantly related to her brothers (1/4): her average relatedness to her sibs is 1/2, just as it would be in a diploid species. If she is to benefit by caring for her sibs, a female must be able to distinguish females from males, and preferentially to raise the former. It seems implausible that, at the origin of sociality, a female should have this ability.

In fact, things are more difficult than this argument suggests. Even a female that preferentially invests in female siblings is not necessarily favoured selectively over one that raises her own offspring. The point is a subtle one. If there are more females in a population than there are males (strictly, if investment in females is greater), then females are reproductively less valuable than males, so the advantage that a daughter gains by preferentially investing in her sisters disappears. Thus, the first female to stay home and look after her sisters (supposing she could distinguish them, as larvae, from her brothers) would indeed benefit, but once the trait was common in the population, the advantage would disappear.

Fig. 16.3 The relatedness of a worker bee to males in the same colony (from Ratnieks, 1988). In (a), the queen mates with many males: in (b), she mates with only one male. We are interested in the relatedness of worker A to unfertilized male eggs laid in the colony: that is, in the fraction of genes in the eggs that are identical by descent to genes in A. Such genes are shown in black: note that males are haploid, and pass all their genes to their daughters, whereas females are diploid, and pass half their genes to their offspring. In (a), the relatedness of worker A to male eggs (B) laid by the queen is $\frac{1}{4}$, and to eggs (C) laid by other workers is $\frac{1}{8}$. It therefore pays workers to destroy eggs laid by other workers. In (b), the relatedness of worker A is $\frac{1}{4}$ to B, and $\frac{3}{8}$ to C. It therefore pays a worker to raise eggs laid by other workers.

This has led people to seek the relevance of haplodiploidy to the origin of sociality in 'split sex ratios' (Grafen, 1986): that is, to situations in which broods produced at different times of year (Seger, 1983) or in different circumstances (Frank & Crespi, 1989) have different sex ratios. If this was true of an initially solitary species, then there would be selection favouring females that could respond to the time or circumstance when broods were female-biased by remaining to help their mothers. Note that this does not require females to be able to recognize the sex of individual larvae.

Split sex ratios may have played a part in the origin of eusociality in the Hymenoptera: certainly, it is hard to see how a causal connection between haplodiploidy and sociality could arise except through split sex ratios. In established eusocial insects, haplodiploidy leads to a conflict of interest between queen and workers over the sex ratio. Trivers & Hare (1976) pointed out that the queen will favour a 1:1 ratio of investment, and workers a 3:1 ratio of investment in females relative to males. They thought that the data support the latter ratio, and hence suggest that the workers control the sex ratio: later work tends to support them.

Termites and cyclical inbreeding

Termites are eusocial but not haplodiploid. There are features of their ecology that favour sociality. They feed on rotting wood, and digest cellulose with the help of gut symbionts. These they must acquire as larvae, and sometimes after every moult, from other individuals. Thus the concentrated nature of their food source, and the need for contact, favour sociality. What of their genetics? One possibility is as follows. In termite (but not hymenopteran) colonies, there are accessory reproductives, descended from the founders, which take over if the original founders are lost. Consequently, the resulting alates (dispersing reproductives) may be somewhat inbred. Mating, however, is between alates from different colonies, so that the first generation of workers in a colony are heterozygous, but genetically similar to one another. Hence these workers are more closely related to one another than to their own offspring. It is hard to say whether this is relevant to the origin of sociality in termites because they have no surviving solitary or primitively social relatives.

Can social insects recognize kin?

Many eusocial hymenopterans can recognize nest-mates. Buckle & Greenberg (1981) investigated this ability in the halictine bee *Lasioglossum zephyrum*, taking advantage of the existence of genetically uniform but unrelated laboratory strains. In this species, the nest entrance is guarded by a single bee: other bees can enter only if the guard permits. Bees of strain X admitted bees of that strain, whether from the same or a different nest, but not bees of strain Y. An artificial nest was constructed, containing one bee of strain Y, and the remainder of strain

X. If the entrance was guarded by one of the X-strain bees, either X or Y bees were admitted: if the entrance was guarded by the single Y-strain bee, only X-strain bees were admitted. In other words, a bee learns the genotypes of other bees in its colony, and accepts them as nest-mates, but does not know its own genotype.

This capacity is of obvious importance in ensuring cooperation between nest-mates. There is no convincing evidence that eusocial insects can distinguish lineages within a colony, although such an ability could be selectively advantageous—for example, in deciding whether to invest in juveniles or to cull them. In other words, the ability of bees to recognize others by their genotype is of a kind that helps the survival of the colony, but not of particular lineages within the colony: this conclusion may well be altered by future research.

16.5 The division of labour in animal societies

We have already met examples of increased efficiency arising from the division of labour: for example, through specialized enzymes (Chapter 7), and through the differentiation of germ line and soma in the Volvocales (Chapter 12). The same principle is illustrated by insect castes: the account that follows relies heavily on Wilson (1975) and Page & Robinson (1991).

By definition, all eusocial insects show a division of labour between reproductive and housekeeping/foraging roles. In many, there is a division of labour between workers as well. The simplest type of specialization is 'age polyethism': workers of different ages perform different tasks. In honey-bees, the following age castes are currently recognized:

- Workers from 0 to 2 days old clean cells.
- Workers from 2 to 11 days old care for larvae and the queen.
- Workers from 11 to 20 days old process the incoming food.
- Workers over 20 days of age collect pollen and/or nectar for 1–3 weeks, and then die.

Thus the same bee performs different tasks as it matures. Functioning of the whole colony depends on the fact that several generations of workers live together. The division of labour is not rigid, since colonies can cope with variations in demography and food availability. In part, this flexibility is achieved by atypical age-dependent behaviour. There can be accelerated or arrested worker development (precocious foragers and over-age nurses), and behavioural reversion (e.g. from foraging to brood care).

How is the division of labour, and behavioural plasticity, regulated within the colony? A major physiological mechanism is the regulation by neurosecretory brain cells of the secretion of juvenile hormone by the corpora allata. The titre of

juvenile hormone increases as the adult bee ages. Bees of a given age treated with different quantities of juvenile hormone perform different tasks. Robinson (1987) proposed the following model. Juvenile hormone titre increases with age by a genetically programmed process, and influences the response thresholds of the central nervous system to task-associated stimuli. Thus, young bees would have a low threshold for brood care and a high one for foraging. The strengths of these stimuli depend on conditions in the colony, which can also affect juvenile hormone titre. Such a mechanism can ensure that individual task performance adapts to colony needs.

In ants and termites, the different castes are analogous to the differentiated cells of an organism: they do not assume different roles as they age.

Why is caste specialization advantageous? Note that it is not sufficient that some workers nurse and others forage. The requirement is that two task generalists performing tasks 1 and 2 should be less efficient than two specialists, one performing each task, acting together. This will be so if different tasks require different specialized organs, whose maintenance is costly. The generalist either has full-blown specializations, which are uselessly costly half the time, or has inferior tools. There is indeed evidence that bees have age-related special tools. When bees shift from brood care to foraging, the hypopharyngeal gland, which produces food for the brood, degenerates. This change can be elicited by a large dose of juvenile hormone. To give a second example, Withers *et al.* (1993) found that there are age-related changes in the structure of a region of the brain concerned with sensory integration and learning. Specialized tools are more obvious in many ant species, in which worker castes differ dramatically in size and weapons.

Since reproductive castes can 'reproduce' only through queens and males, it is reasonable to assume that the mix of castes is optimal for the production of queens and males. Figure 16.4 shows a simple model of this optimization process. The model can easily be extended to show:

- If there are two castes but only one task, the less-efficient caste will be eliminated.

- Stable coexistence of two castes requires that there be at least two tasks, and that each caste is more efficient than the other at performing one of the tasks.

- Evolution will permit the establishment of new castes, until their number equals the number of distinct tasks.

- As the efficiency of a given caste increases, its biomass in the colony will decrease (contrary to what a simplistic model of individual selection might suggest).

There is an obvious parallel between these conclusions, and those concerning the coexistence of competing species on a range of limiting resources.

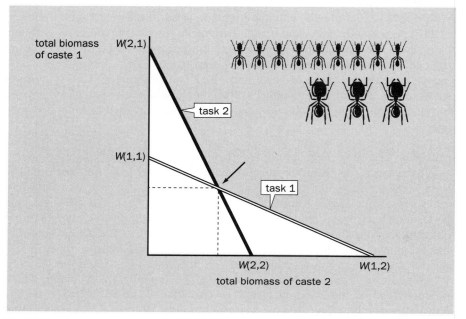

Fig. 16.4 The optimal mix of castes in an insect colony (after Wilson, 1975). Survival of the colony requires that two tasks be adequately performed. Task 1 can be performed by a biomass $W(1,1)$ of caste 1, or $W(1,2)$ of caste 2, or by a linear combination, as shown. Since $W(1,1) < W(1,2)$, caste 1 is specialized to perform task 1. In performing task 2, caste 2 is more efficient: $W(2,2) < W(2,1)$. The minimum biomass of workers able to perform both tasks occurs at the intersection of the two lines.

16.6 Factors predisposing insects to sociality

We have already discussed whether haplodiploidy predisposes the Hymenoptera to sociality: the topic is difficult, but a positive answer is at least plausible. However, haplodiploidy cannot be a sufficient cause: many hymenopterans are solitary. Ecological factors were presumably crucial: if k is large, r need not be. Among these, the existence of a nest, in which the young are provisioned and protected, is perhaps the most important. Parasitoids are a major cause of death in juvenile insects: if two females cooperate, one can protect the nest while the other collects food. A second factor is the existence of a concentrated and defensible resource, either pre-existing (e.g. a log of rotting wood) or constructed (e.g. the massive communal web of social spiders). But as yet we cannot specify precisely the necessary and sufficient conditions for the evolution of sociality.

16.7 The origins of human society

A simple model

The peculiar thing about human society, as Gellner (1988) remarks, is that there is not one kind of society, but many. The problem for biologists, therefore, is to explain the evolution of those traits that make it possible for humans to live in a vast array of different societies. In this section, we describe a simple model—called, for obvious reasons, the Social Contract game—that explains how a group of rational and communicating organisms can defeat the Prisoner's Dilemma. We do not regard this model as an adequate description of any human society. We use it for two purposes. First, it shows that rational beings can, at least in principle, defeat the Prisoner's Dilemma. Second, it provides a background for our later discussion of the fact that real societies are not exclusively, or even mainly, rational.

The model (Maynard Smith, 1983b) is shown in Box 16C. Logically, the contract is stable against defection. Although not an adequate model of any complete society, it is for some groups within society, such as the freemasons or the mafia. Clearly, the behaviour requires language, because the word

BOX 16C The Social Contract game

Individuals have two possible strategies, cooperate, C, and defect, D. The pay-off to an individual depends on what others in the population are doing, as follows:

		rest of population	
		C	D
individual	C	30	10
	D	40	20

This is a typical Prisoner's Dilemma game: it pays to defect, whatever others are doing. The members of the population therefore bind themselves to the contract 'I will cooperate: I will join in punishing anyone who defects'. The cost of being punished is -50, and of joining in punishing is -5. Supposing that all other individuals keep to the contract, the pay-offs to an individual are as follows:

cooperate, and join in punishing	$30 - 5 = 25$
defect	$40 - 50 = -10$
cooperate, but do not punish ('free-rider')	30

Thus the contract is still not stable, it can be invaded by free-riders. It must be amended to read 'I will cooperate: I will join in punishing defectors: I will treat anyone who does not join in punishing as a defector'.

'cooperate' in the contract refers to some specific behaviour (e.g. not smoking in public buildings) that is not innate, but must be defined linguistically. The process whereby the contract is reached also requires a developed form of self-consciousness. Individuals must not only be concerned with their own future interests but they must recognize others as being individuals like themselves, with similar motivations: otherwise, why should one individual suppose he could persuade another to agree to the contract? Only organisms with language, and this level of self-consciousness, would be capable of negotiation leading to a contract.

The model has two major weaknesses. The first concerns the way in which cooperation is enforced, and the second the assumption that all individuals start out equal. First, consider enforcement. It is probably not true that most social behaviour is motivated by a fear of punishment. Often, the crucial factor is ritual. In Gellner's (1988) words:

> The way in which you restrain people from doing a wide variety of things, not compatible with the social order of which they are members, is that you subject them to ritual. The process is simple: you make them dance round a totem pole until they are wild with excitement, and become jellies in the hysteria of collective frenzy: you enhance their emotional state by any device, by all the locally available audio-visual aids, drugs, dance, music, and so on: and once they are really high, you stamp upon their minds the type of concept or notion to which they subsequently become enslaved.

It seems, then, that biologists have to explain not only the origin of language but also the capacity to be socialized by ritual. The content of ritual varies from society to society, and its importance relative to rational calculation and legal sanction may have diminished in the course of history (or perhaps we only hope that it will diminish), but all humans are influenced by it. This may be no great mystery. An individual who could not learn would be ostracized. But it is still unclear why the inculcation of proper behaviour is achieved by ritual and myth rather than by explicit precepts. Why Macbeth, and the biblical story of Cain and Abel, instead of, or as well as, the simple rule 'thou shalt not kill'? It is easy to say that ritual is effective in creating an emotional commitment to a set of beliefs, rather than a mere acceptance, but that is to describe a state of affairs rather than to explain it.

The innate capacity to be influenced by ritual may have been individually selected, although we are not clear why it should be so. But rituals, once formed, are culturally, not genetically, transmitted. Boyd & Richerson (1985) have suggested that between-group selection may have favoured rituals that are particularly effective in binding a group together. They are not suggesting that the content of a myth or ritual is genetically specified. Their point is, rather, that a culturally transmitted trait can also be selected. If a human group is successful because of its system of ritual, this has two effects: by cultural evolution, it causes the spread of a particular set of beliefs, and by genetic selection it favours

individuals who can be strongly influenced by those beliefs (and probably by any other ritually enforced beliefs). In other words, there is between-group selection for culturally inherited systems of belief that favour the success of groups, and there is individual selection for the genetically inherited ability to be influenced by ritual.

A second weakness of the model lies in the assumption that all individuals are initially equal, and have the same strategy set open to them. In more complex societies this is clearly not so. Possession of land, factories or weapons alters the options available. The Marxist insistence that political ideology is determined by class position is clearly part of the truth. It is also easy to understand why social groups should acquire systems of myth and ritual: a group with such a system will be better able to defend the interests of its members. The persistence of historically shared myths is astonishing, and sometimes disastrous. Such group identities, within a larger community, are more relevant to the troubles of existing societies than to their origins, but they do bear witness to the ease with which humans construct group identities, buttressed by ritual.

Theories of society

Theories of human society differ in two main ways:

- Does society resemble a house or a termite mound? A termite mound differs from a house in that no individual has an image of the final structure, which, although highly functional, emerges from the interactions of millions of individuals, whose behaviour is law-governed but not influenced by any such image. In contrast, an architect starts with an image of the final building, which is functional because of his rational thought, and not through the naturally selected but blind behaviour of the builders. Does society resemble a house or a termite mound? Was its structure planned by a group of rational law-givers, as the constitution of the United States was planned by the founding fathers? Or is it merely the unplanned outcome of the behaviour, rational or otherwise, of its members?

- Is individual behaviour determined by reason or ritual?

These two dichotomies define four theories of society, which we now describe. We have named them after four famous philosophers. No doubt we shall be accused of grossly oversimplifying their ideas. This is perhaps true, but it may be no bad thing. The trouble with theories of society is that their formulations are so lengthy and so complex that they are hard either to grasp or to test. Theories should be formulated briefly, even if the facts to which they are relevant must be described at length. *The Origin of Species* is a long book, but no biologist regards it as impossible to outline Darwin's theory in a few sentences. It is true that some mathematical theorems have proofs running to several hundred pages: in essence, the latter constitute theories associated with the theorems. But mathe-

matical statements are unambiguous, and so permit long chains of argument. Unfortunately, statements about society, as about evolution, have a degree of ambiguity. It follows that theories in these fields, if they are to be operative in the sense of leading to clear predictions, must be simple. Of course, it may be that no operative theories are possible in the social sciences. With this preliminary excuse, we offer the following classification:

Rousseau and the social contract. This is the model we have already discussed. Its essential features are that society is rationally planned, and that its maintenance depends on rational behaviour by its members.

Plato's republic. Plato's model of society is also a house rather than a termite mound. It is planned by a philosopher. It depends on a division between rulers, soldiers and labourers, which, for some odd reason, Plato calls 'justice'. But it is maintained by myth, not by the rational calculation of its members. With admirable frankness, Plato wrote in the *Republic*

> How can we contrive one of those expedient falsehoods we were speaking of just now, one noble falsehood, which we may persuade the whole community, including the Rulers themselves, if possible, to accept? . . . All of you, we shall tell them, are brothers, but when God was fashioning those of you who are fit to rule, he mixed in some gold, so these are the most valuable, and he put silver in the auxiliaries, and iron and bronze in the farmers and other craftsmen. Since you are all akin, your children will mostly be like their parents . . . That is the story. Can you suggest any device by which we can get them to believe it?

In case there should be any doubt that it was ritual that should be used to ensure compliance, he wrote in the *Laws*:

> All he [i.e. the legislator] needs to do is to find out what belief is most beneficial to the state, and then use all the resources at his command to ensure that throughout their lives, in speech, story and song, the people all sing to the same tune.

Durkheim's organic society. Durkheim's society is a termite mound. It is functional, but no-one planned it. It works because its members are influenced by ritual.

Adam Smith and the free market. Strictly, Smith's model is of the economy, not of the whole society, but perhaps, since his most famous recent practitioner announced 'there is no such thing as society', that need not worry us. His basic idea is that the economy works best, in producing the goods people want, if it is free rather than planned. Individuals do behave rationally, but in their own individual interests, with no image of the total result. It is a termite mound, not a house, but one maintained by rational calculation. The history of the last 40 years leave little doubt that he was right about the relative efficiencies of free and planned economies. The snag is that his truth is partial. It is profitable to discharge toxic wastes into rivers, and to market cigarettes and heroin.

Not all theories of society can so easily be forced into this Procrustean matrix. Kant's views are summarized in the following quotation:

The problem of organizing a state, however hard it may seem, can be solved even for a race of devils, if only they are intelligent. The problem is: given a multitude of rational beings requiring universal laws for their preservation, but each of whom is secretly inclined to exempt himself from them, to establish a constitution in such a way that, although their private intentions conflict, they check each other, with the result that their public conduct is the same as if they had no such intentions.

In other words, we need an architect who will so design the rules that individuals, each acting in their own best interests, will generate a socially desirable end. Society is a termite mound, in that its members do not envision the whole, but their behaviour has been planned by a designer so that they will generate the society he already envisions.

Thus in Kant's model, as in Rousseau's (at least, in the Social Contract model described above), individual behaviour is governed by rational self-interest. In both, there is also a need for a central planner, but in Kant's model, instead of laying down specific rules of behaviour, with punishment for transgressors, the central plan is confined to laying down the rules of interaction in such a way that rational self-interested behaviour leads to the common good. Adam Smith could have claimed that this is what he did, at least for the economy, by insisting that the market should be free, and not constrained.

Marx and Engels were, in this sense, Kantian. Marx's thesis 'man's being determines his consciousness' implies that an individual's behaviour is influenced by his own needs: it is a termite-mound model. But they argued that, if land and the means of production are publicly owned, it will pay individuals to act in a socially desirable way. The snag was that they did not allow for the free-rider, or did not foresee that state machinery intended to discipline the free-rider would develop into a new social class, governing in its own interests.

The ideas of Adam Smith and Karl Marx may be more relevant to our present political problems than to the origins of society. The Durkheimian view of a functional but unplanned society, governed by ritual, is closer to the picture of the earliest societies held by most anthropologists. Such a society requires individuals genetically capable of language, and of socialization by ritual. But this organic picture cannot be the whole story. Organisms function effectively only because natural selection has moulded the rules governing the behaviour of cells so as to produce a functional whole. How do ritual-governed societies become functional? After all, it is probably as easy to inculcate by ritual behaviours that are socially destructive. There seem to be two possible answers. One is that early societies became functional through the selective success of culturally inherited sets of beliefs. The alternative is that an element of rational calculation played a role in deciding which particular beliefs and practices would be enforced by ritual. Plato's legislator, inventing rituals to enforce his noble lie, is going too far,

but those who elaborated myths and rituals may have given some thought to the social consequences of their inventions.

Origins

What can be deduced about the origins of human society from the fossil record and from comparative biology? Figure 16.5 summarizes hominid fossil history. Our australopithecine ancestors adopted an upright posture at least 4 million years ago, long before there was any substantial increase in brain size. Australopithecines had ape-sized brains, about one-third the size of a modern human's. By the *Homo erectus* stage, the brain had doubled in size. Even so, the tools of *Homo erectus* were technically uninventive. Their most elaborate tool was the handaxe, fashioned from a single block of stone and worked on both surfaces. It was more innovative than any tool used by chimpanzees. Chimpanzees use hammerstones and anvils in the wild in some habitats: in captivity, they can be taught to make and use flint flakes.

It is a striking fact that handaxes, which first appeared about 1.5 million years ago, continued to be made for more than a million years. Only after the evolution of the earliest *Homo sapiens*, around 250 000 years ago, are there signs of slightly more skilled toolmaking and a more varied toolkit. However, tools and tool materials generally remained conservative, even after the appearance of fully modern, large-brained people around 100 000 years ago. The burst of technological innovation came relatively recently, only about 40 000 years ago.

It is hard to suppose, therefore, that the increase in brain size, by a factor of almost three, could have been a response to selection for improved technical skill. What selective force did lead to our larger brains? It is conceivable that the relevant factor was the evolution of language. It seems more likely, however, that language as we now know it evolved rather recently, and that it was responsible for the dramatic changes that have occurred in the past 100 000 years, not for the increase in brain size that took place earlier. Dunbar (1992) has pointed out that, if one compares the forebrains of existing primates, the best predictor of brain development is the size of the social group in which an individual lives. He therefore suggests that the main selective force favouring increased intelligence in primates arises from social interactions. An individual who can act appropriately in a variety of social contexts will be fitter than one who cannot. Extending this argument to our own ancestors, the increase in brain size was an adaptation to living in society.

Half a million years ago, *Homo erectus* was distributed throughout the tropical and temperate regions of the Old World. This has led to a debate between those who think that existing human races evolved *in situ* from the local populations of *H. erectus*, and those who think that *H. sapiens* originated once only, probably in Africa (although that is not certain), and subsequently spread round the world, replacing the local populations of *H. erectus*. The latter view, of a single origin,

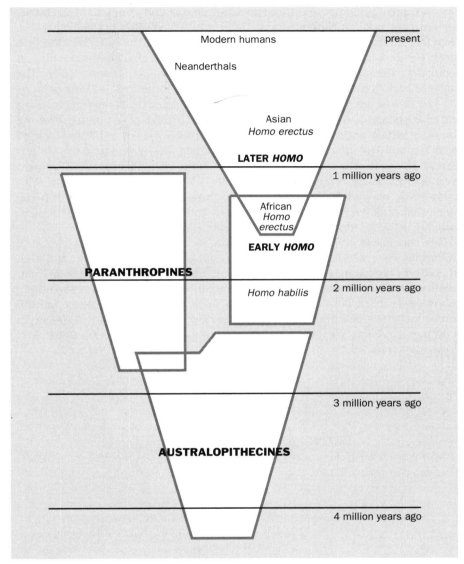

Fig. 16.5 The human fossil record (after Stringer & Gamble, 1993). Four grades are tentatively recognized. Australopithecines were early, small-brained but bipedal hominids. There was then a division into two main ecological types—heavily built and mainly vegetarian paranthropines, and a more slenderly built lineage, *Homo*, showing an increase in brain size, leading to Neanderthals and modern humans. This distinction between 'gracile' and 'robust' forms already existed among the later australopithecines. Early *Homo* and *Paranthropus* co-existed for at least 1 million years. There is still debate about the number of co-existing species within each lineage.

has been greatly strengthened by molecular data (Cann *et al.*, 1987). The argument goes as follows. Mitochondrial DNA is inherited solely in the female line. Therefore all existing mitochondria are descended from the mitochondria of a single female living at some unknown time in the past: the smaller the human population, the more recent we would expect that common ancestor to be. The range of variation of existing mitochondria can be measured, and an estimate can be made of the rate at which variation accumulates (by comparing humans and chimpanzees). Given these two measures, Cann *et al.* estimated that the common female ancestor lived some 200 000 years ago. It is also possible to make a rough estimate of the human population size, given that this date is correct: the answer is of the order of 10 000. If the estimate of 200 000 years is even approximately correct, it rules out the idea that human races were independently derived from local *H. erectus* populations. We are left with the conclusion that we are descended from a rather small human population, probably living in Africa some 200 000 years ago.

This conclusion from the molecular data is consistent with the fossil and archaeological evidence. By 40 000 years ago, populations in Europe and elsewhere were producing a wide range of novel artefacts, were burying their dead, painting on the walls of caves, and engaging in trade. It seems that something happened to make possible both the geographic spread of *H. sapiens*, and the burst of new technical and cultural practices. The obvious candidate is the origin of human language. We discuss this, the last of the major transitions, in the next chapter.

17 The Origin of Language

17.1 Introduction

The past 30 years has witnessed a debate between the holders of two very different views about how humans are able to talk. The behaviourists, following B. F. Skinner, argue that we learn to talk in the same way that we learn any other skill. Children are rewarded when they speak correctly, and reproved when they make mistakes. We can talk, whereas chimpanzees cannot, because we are better at learning: there is nothing special about language.

In contrast, Noam Chomsky and his followers have argued that humans have a peculiar competence for language, which is not merely an aspect of their general intelligence. We learn to utter, and to understand, an indefinitely large number of grammatical sentences, and to avoid an even larger number of ungrammatical ones, so we cannot possibly learn which sentences are grammatical by trial and error. Instead, we must learn the rules that generate grammatical sentences. These rules are of great subtlety, so that, although we acquire and apply them, we cannot formulate them explicitly. For example, consider the two following sentences:

How do you know who he saw? (1)

Who do you know how he saw? (2)

Every speaker of English knows at once that (1) is grammatical, and (2) is not. But what rule tells us this? No-one but a trained linguist would have any idea, any more than a non-biologist would know how the rate of beating of the heart is adjusted to meet changing demands. In section 17.3, we describe a hypothesis about the rule that tells us that (2) is ungrammatical: it is a subtle rule, but as yet no-one has thought up a simpler one. It is hard to believe that we could so painlessly master such rules unless we were genetically predisposed to do so. More generally, it is still beyond the wit of linguists and computer scientists to write a language-translating programme, yet many 5-year-olds know two languages, do not mix them up, and can translate from one to the other.

A second reason for thinking that we cannot learn to talk by trial and error lies in the poverty of the input on which a child must rely. After hearing a finite set of utterances, a child learns to generate an indefinitely large number of grammatical sentences. This implies that the child learns rules, and not merely a set of sentences. As we have just seen, the rules are of great subtlety. To make matters worse, many of the sentences that a young child hears are not in fact grammatical, and parents usually do not correct a child's ungrammatical utterances. In section 17.8 we give a third and decisive reason for thinking that the ability to learn to talk is independent of general intelligence.

We are satisfied that the debate between the behaviourists and the proponents of a special linguistic competence has been won by the latter. But it is hard to say what it is that is innate. It is not the ability to speak a particular language, but the ability to learn *any* language. In communication, meaning is converted into

a linear sequence of sounds (or gestures, in deaf-and-dumb language): Chomsky has referred to this sequence as the 'surface structure'. It is perhaps worth pointing out that linearity is not a necessary feature of a communication system: our messages could be multichannel, as imagined by Harrison (1985), or could consist of two-dimensional displays. But before we can communicate, we have to have something to say. In order to think, we must be able to represent the world in our minds, and to manipulate the image we have formed: thinking requires a kind of mental grammar. The fact that the Greeks used the word *logos* to mean both discourse and reason underlines the close relationship between the two processes. In section 17.2, we argue that the evolution of linguistic competence required the acquisition of two abilities. The first, required both for language and for thinking, is the ability to form complex representations of the world in our minds. The second, more specific to communication, is the ability to transform these representations into a linear sequence of symbols.

In section 17.3, we turn to the problem of syntax—of how our meaning can be converted into a sequence of sounds. We cannot hope to give a complete account of modern generative grammar, but we try to give a taste of what is involved, because we must first understand human grammar to grasp the difference between animals and humans that had to be bridged in evolution. In section 17.4, we discuss briefly how children learn to talk. We show by an example that, in learning to talk, children do not follow a rule of thumb, but learn the grammatical structure of a sentence.

In section 17.5, we argue that linguistic competence, like any other complex and adapted organ, has evolved by natural selection, and consider briefly the objections to this idea that have been raised by linguists. In section 17.6, we draw an analogy between the development of language and of tool use, by comparing the abilities of animals and of children of different ages. It seems that there is not only a formal similarity between the construction of sentences and the performance of manual tasks, but there may be a common physiological basis for the two abilities. In particular, there is a hint that linguistic training improves manipulative abilities. There may, therefore, have been mutual reinforcement between tool use and language among our ancestors.

In sections 17.7 and 17.8, we describe impairments of linguistic ability arising from physical injury, and genetic change, respectively. The genetic data are particularly exciting, since they offer the hope that we may one day be able to dissect linguistic competence, as we are already dissecting embryological development.

In section 17.9, we describe 'protolanguage', as observed in animals, very young children, humans who have been deprived of linguistic input, and the 'pidgins' used between adults who have no common language. In section 17.10, we discuss the transition from such protolanguage to fully developed language, and ask whether any functional intermediates could have existed between them.

There are two striking analogies between the role of language in the origin of

human society, and earlier evolutionary transitions. The first was pointed out by Jablonka (unpublished work, 1994). The origin of multicellular organisms required a second inheritance system, based on methylation and other means of labelling DNA. This made a division of labour between cells possible, and may have helped to bind groups of cells together, so that the organism rather than the single cell became the target of selection. In human society, language constitutes the second inheritance system. It makes a division of labour possible, and, through myth and ritual, binds human groups together.

The second analogy may be even more fundamental. Chomsky has repeatedly emphasized that the essential feature of language is that one can utter an indefinitely large number of meaningful sentences. In Chapter 3, we argued that the origin of life required replicators that could exist and be copied in an indefinitely large number of forms.

Finally, we must emphasize that neither of the authors is trained in the subject matter of this chapter. We have relied heavily on two sources: Bickerton's (1990) book, and the review by Pinker & Bloom (1990).

17.2 Language and representation

The East African vervet monkey has three alarm calls: one for pythons, one for martial eagles and one for leopards. A call is uttered whenever a monkey sees one of these animals, and triggers an appropriate response in those monkeys that hear it: for example, the leopard call causes the monkeys to climb higher into a tree, a response that would be unwise if the predator was an eagle. It has been suggested that language can be traced back to such an alarm system. This implies that the alarm signal is in effect a word, such as *python*: that is, the call is a sign for a python.

Is this so? It could be that an alarm call reflects only a degree of danger, and that a monkey hearing the call looks round, sees a python or a leopard, and reacts accordingly. Seyfarth & Cheney (1992) ruled out this explanation by showing that monkeys react to recorded calls in exactly the same manner.

The acid test of the semantic nature of these calls relied on habituation studies. If an alarm signal is not followed by an objective threat, animals become used to it and cease to react. The calls *wrr* and *chutter* are both used by vervets to call attention to another group of monkeys. If a vervet is habituated to one of these calls, will it transfer the habituation to the other? If it does, this would be evidence for semantic representation: the transfer requires that the monkey interprets *wrr* as a sign for another group of monkeys, and interprets *chutter* in the same way.

The answer to the question is complicated by the fact that vervets, like many other mammals and birds, can identify individuals. If habituation to the *wrr* of one particular monkey has occurred, it is transferred to the *chutter* of the same

individual, but not to the *chutter* of others. Seyfarth & Cheney reasonably conclude that these calls are semantic representations. But word usage in vervets is severely limited: in particular, a vervet does not use a call to refer to an object that is not present here and now (Premack, 1985).

Bertrand Russell argued that there is a mapping between objects 'out there' and words, or alarm calls in our example. The snag is that the mapping is not direct. First, there is a mapping from the object python to the concept 'python', and then from 'python' to the word *python*, as de Saussure was already aware. Vervets must have a concept of python in their brains, otherwise they could not recognize them. In fact, young vervets have to learn the semantic range of their calls, just as young children must learn the semantic range of words.

Humans use words such as 'unicorn' and 'golden mountain' to refer to objects that do not exist in the world, and for which there is therefore no sensory evidence. No animal would have an alarm call for a unicorn. It is true, however, that vervets use alarm calls as a means of deception—for example, to drive other monkeys away from a food item. But this is not a very shrewd deception: the deceiving monkey shows no signs of panic, even if others can clearly see what he is doing. Also, animals have calls only for objects of immediate biological interest, whereas humans can communicate about anything. For us, a potentially complete representation system was decisive.

An important analogy emerges here. At the beginning of the book, we observed that the origin of life had been marked by the passage from limited to unlimited hereditary replicators. It now seems that the origin of human life was marked by the passage from limited to unlimited semantic representation.

Humans use language for communication, but it may well be that the most important aspect of language is that it is used for internal representation in the brain. Pre-linguistic animals can have a representational system. For example, the sea anemone *Stomphia coccinea* may encounter 11 species of starfish, of which two are its predators. If an individual from a non-dangerous species touches the sea anemone, nothing happens. If a predatory individual touches it, however, it immediately withdraws. This requires that the sea anemone can distinguish between different starfish species: there is a link between information and behaviour (Ward, 1962).

Representations invariably lead to the formation of categories, such as dangerous and harmless starfish. Category formation is the grouping together of representations that require similar behavioural responses. This is independent of whether or not the representations are innate. A category is a tacit concept: that is why animals must be granted the faculty of concept formation (Bickerton, 1990). Language has merely provided labels for concepts derived from pre-linguistic experience.

Categories are likely to be arranged in a hierarchical fashion, particularly if they are associated with words. All lexical items are part of a hierarchical structure (Fig. 17.1). When we are asked to say what a word means, we

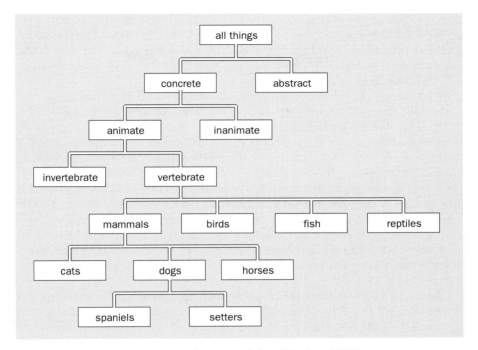

Fig. 17.1 The hierarchical nature of concepts (after Bickerton, 1990).

usually answer in terms of such a hierarchy: 'a spaniel is a kind of dog used in hunting'.

A feature fundamental to language is the subject–predicate distinction. For example, we do not have four separate concepts, and corresponding words: 'dog-running', 'dog-sleeping', 'lion-running', and 'lion-sleeping'. Instead, we have two nouns, *dog* and *lion*, and two verbs, *run* and *sleep*. To say 'the dog is sleeping' is a predication. Since many qualities can be predicated of each entity, the range of things that can be said with a vocabulary of a given size is far greater than it would be if we had a separate word for every conjoined pair, entity + quality. The subject–predicate distinction is a universal feature of language, and is the basis of our ability to say, and to understand, an indefinitely large number of sentences. But it is also a valuable aid to thought.

Human thinking requires more than the formation of hierarchically arranged categories, and predication. It also requires that we should form structured representations of the world, which can then be manipulated in our minds. For example, we can think 'Two leopards climbed into that tree yesterday: only one has come down, so there is still one leopard in the tree: if I walked under the tree, it might attack me'. It would be selectively advantageous to think such thoughts,

even if we could not communicate them. Such thinking requires, not only that we have noun concepts (leopard, tree) and verb concepts (climb, attack), but that we can arrange these concepts in a pattern, and then alter the pattern. The latter ability is analogous to grammar, as is clear from the use of the words *so*, *if*, *into* and *under* in describing the thought.

In fact, such purely grammatical words such as *so* and *if*, having no demonstrable referent, account for about half the words in human speech. Also included are inflections like *-ing* and *-ed*. Nouns have to be singular or plural, most verbs must indicate the present or past tense, and so on. Grammatical items hold the more meaningful parts of sentences together. They indicate relative number, relative time, relative direction, relative familiarity, relative possibility, possession, necessity, existence, and so on. It is remarkable, however, that such grammatical items refer to only a few of the possible relations, although some languages can convey distinctions not possible in English. For example, both Turkish and Hopi have verbal inflections implying that a statement is based on personal experience, or on second-hand information, but no language has a grammatical distinction between edible and inedible, or between hostile and friendly.

Purely grammatical items, then, are needed for speech, but, as we suggested by our example of the two leopards in a tree, the relationships they express are also needed for thought. This can be illustrated by what is perhaps the most non-verbal kind of thinking we ever do—playing chess. A chess player might think 'If I play P × B, he could then play N–B6, forking my K and Q, so I must not take the bishop'. When thinking, the nouns and verbs would be replaced by visual images of the chessboard, but the purely grammatical items, *if*, *then* and *so*, would remain. Of course, no hominid ancestor ever left more descendents because he could play chess. The point is that the same processes that are needed for speech—concept formation, predication, and the recognition of relationships between concepts—are needed also for thought. The necessary mental abilities may have evolved in the first instance for thinking, rather than communication, or at least for thinking as well as communication.

Communication, however, requires not only an adequate representation of the world in the brain. It requires also that this representation be converted into a linear sequence of sounds—or, in reception, that a linear sequence of sounds be converted into a representation in the brain. This is the topic of the next section. We have not the space to give a full account of generative grammar, even if we were competent to do so, but we hope to give a flavour of the ideas involved.

17.3 Some features of syntax

The Oxford English Dictionary defines syntax as 'the order of words in which they convey meaning collectively by their connection and relation'. Syntax is regarded by many as the key component of any language. Take any ordinary

sentence of 10 words. The words can be arranged sequentially in 3 628 800 different ways, only a few of which are grammatical. There is no way one could learn to avoid the non-grammatical sentences by negative examples: rather, we must process some recipe for constructing sentences.

The modern view of syntax and the innateness of a universal grammar is largely due to Chomsky (1957, 1968, 1975) and his associates. He was the first to distinguish systematically between semantically nonsensical sentences, such as *colourless green ideas sleep furiously* (as opposed to *revolutionary new ideas occur infrequently*), and syntactically wrong sentences such as *new occur revolutionary infrequently ideas*. We present an incomplete account of generative grammar, following the accounts of Bickerton (1990) and Aoun (1992). The reader has to be warned that this may be heavy going.

Consider the following sentences:

The cow ate the grass	(3)
The cow with the crumpled horn ate the grass	(4)

'The cow' in (3) has been replaced by the noun phrase 'The cow with the crumpled horn' in (4). Phrase structure is perhaps the most important element of syntax (Jackendoff, 1977). The structure of this noun phrase is shown in Fig. 17.2, and the general structure of a phrase in Fig. 17.3. A feature of this structure, thought to be universal, is that no more than two branches spring from the same node. In the construction of sentences, phrases can be stacked inside one another like Chinese boxes. An extreme example is the sentence from which (4) is taken: 'This is the man all tattered and torn, who loved the maiden all forlorn, who milked the cow with the crumpled horn, that kicked the dog that chased the cat that killed the rat that ate the malt that lay in the house that Jack built.' The fact that children enjoy such rhymes shows the pleasure they get

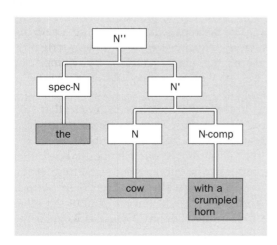

Fig. 17.2 The structure of a noun phrase (after Bickerton, 1990). N (*cow*) is the head of the phrase, and must be a single word. It is first linked to its complement (*with a crumpled horn*) via the node N', and N' is then linked to the specifier (*the*) via the node N'', representing the full phrase. Although the head, N, must be a single word, the complement may itself be a phrase, as it is in this example. Thus phrases are like chinese boxes, stacked inside one another.

Fig. 17.3 Universal phrase structure (after Bickerton, 1990). X represents any lexical category (noun, verb, adjective, preposition, etc.) that can be expanded to form a phrase. The *n* after X′ indicates that this level, unlike X and X′, can be repeated (e.g. *big bad wolf*). The parentheses round the specifier and the complement indicate that these items are not obligatory. The horizontal two-headed arrows indicate that the relative positions of the items can be exchange.

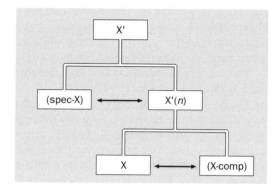

from mastering grammar, even if they would be puzzled to be told that a sentence is a set of hierarchically embedded phrases. Of course, there is a difficulty. It may be possible for linguists to parse sentences in this way, but is there any reason to think that actual speakers do the same, albeit unconsciously? One such reason is given below, in section 17.5: it is an example of a more general argument—that only by recognizing phrase structure are we able to tell grammatical sentences from ungrammatical ones.

A second feature of grammar concerns the 'argument structure' of verbs. Consider the following sentences;

<div align="center">

John sleeps. Mary runs. (5)

The dog bit John. John ate the pie. (6)

John gave Mary a present. (7)

</div>

These sentences illustrate the fact that some verbs (*sleep, run*) have only an 'agent'; other verbs (*bite, eat*) have an agent and a 'patient'; still others (*give*) have an agent, a patient and a 'goal'. This is expressed by saying that a verb has one, two or three arguments. All languages have a similar argument structure. When learning a new language, we do not make mistakes; for example, by trying to give *sleep* a patient. This has led linguists to say that a knowledge of argument structure is innate.

It is worth asking what is 'innate' in this case. It cannot be innate that *sleep* has one argument and *bite* has two. A child must learn the meaning of these words—that is, the actions in the real world to which these words refer. It is a fact about the world, not about either grammar or our genetic makeup, that when you bite, you bite something, but that you cannot sleep something. If you have learnt the meaning of *sleep*, and later learn that the French for sleep is *dormir*, you could hardly imagine that *dormir* could have a patient. In other words, all that is innate is the ability to learn how, in any particular language, to convert the concept of a dog biting John into *the dog bit John*.

A third feature is the ability to alter the meaning of a sentence by moving words about. For example, we can convert the statement *John is hungry* into the question *Is John hungry?* by changing the word order. A curious feature of this process that linguists insist upon is that, when words are moved, we keep track of them. For example, in the sentence *What did you see?* there must be a 'null-element', *What did you see —?* The idea is that the verb *see* normally has a patient immediately following it: the word *what* bears the thematic role of patient, but has been moved to the start of the sentence (we cannot say *Did you see what?*). The null-element marks the place from which *what* has been moved. Why on earth should anyone believe this? Certainly, speakers of English are not aware of leaving null-elements behind.

The answer is that we have grammatical skills that can only be explained by null-elements. For example, at the beginning of this chapter we pointed out that English speakers know that

$$\text{How do you know who he saw?} \qquad (1)$$

is a possible question, but

$$\text{Who do you know how he saw?} \qquad (2)$$

is not. The explanation in terms of null-elements goes as follows. The second question should be written *Who do you know how he saw —?*, where — marks the place from which the object of *saw*, now represented by *who*, has been moved. But a constraint on movement states that a *wh*-word cannot be moved across a space occupied by *how*. Knowing this, we know that (2) is ungrammatical.

It is hard for an outsider to decide how plausible this argument really is. We know that (1) is grammatical, but (2) is not. The linguists' argument is that the easiest way of explaining this grammatical insight (and, of course, others) is by postulating null-elements. As biologists know to their cost, there are dangers to this plausibility argement. There may be a simpler way we have not thought of, or natural selection may not have picked on the simplest way. For example, the simplest way of making segments is probably the one suggested by Turing (section 14.2), but it seems that animals do not do it that way. However, the reason we have spent time on null-elements is to show how subtle the grammatical rules may be: in speaking, just as in digesting or making segments, we follow complex rules of which we are not conscious.

17.4 Language acquisition

The main reason for thinking that humans have an innate competence for learning language is that children acquire a mastery of complex grammatical rule by hearing others talk. We describe one example to illustrate this point (Crain, 1991). How do children learn to construct Yes/No questions? Consider

the two following transformations:

John is tall. Is John tall? (8)

Mary can sing well. Can Mary sing well? (9)

Hearing such sentences, the simplest rule that a child could learn in order to construct Yes/No sentences would be: 'Move the first *is* (or *can*, *will*, etc.) to the front of the sentence'. Unfortunately, the rule fails in the following slightly more complex example:

The man who is running is bald. Is the man who running is bald? (10)

The reason why the simple rule fails is that it ignores the phrase structure of the sentence: *the man who is running* is a noun phrase, which cannot be broken up by moving *is* from the middle of it. Only by recognizing this phrase structure can a speaker formulate the correct question:

Is the man who is running—bald? (11)

Crain found that children between 3 and 5 years old, in an experimental situation, never produced questions like (10).

17.5 Natural selection for language

The main architect of the theory of universal grammar, Chomsky, put a strong emphasis on the innateness of our language acquisition device. It would therefore seem natural to assume that an evolutionary origin of language is widely accepted by modern linguists. Yet this is not so. Chomsky once said that, although we do have a 'language organ', to speculate about its origin is as futile as to do the same about any other organ—for example, the heart. This is baffling to an evolutionary biologist, who would make a 180° turn, and argue that one should contemplate the origin of any organ, including our language device.

In this section, we discuss whether the origin of language can be explained by natural selection. Our treatment follows rather closely that of Pinker & Bloom (1990). We start from the presumption that natural selection is the only plausible explanation for adaptive design. What other explanation could there be? Following the famous paper by Gould & Lewontin (1979), one could suppose that language is a spandrel: that is, an unselected byproduct of design for some other purpose. More specifically, language could be a modified or an unmodified spandrel. If the claim is only that language is a modified version of a structure that once served some other function, the answer is an (almost trivial) yes: the claim is true of most complex structures. But if language is modified, then natural selection was the modifying force. The only interesting non-selectionist alternative would be that language is an unmodified spandrel—a perfect pre-adaptation. No serious theory has been built along these lines.

Non-selectionist explanations of this kind only work if you want to use something complex for something simpler. You could use a dead computer as a

paperweight or an aquarium, but an ordinary paperweight or aquarium could never by used as a computer. Since language is a very late and complex evolutionary invention, it cannot be an unmodified spandrel.

There are, we think, mental abilities that are unselected spin-offs of general cognitive ability: one that we have mentioned is the ability to play chess. If we thought that language competence was merely a consequence of a general increase in intelligence, then we would be willing to treat it as a spandrel—that is, a trait whose origin hardly calls for an explanation. But this is precisely what Chomsky and his associates deny: they insist that language is special. If the position we have taken in this chapter is correct, then one aspect of language competence, namely the ability to form and to manipulate mental representations, did evolve because of selection for general cognitive ability: only the specific competence to convert meaning into surface structure evolved because of selection for communicative ability.

It would be hard to argue convincingly that language lacks design for a goal. It solves both the problems of internal representation and of interindividual communication. Grammar maps a hierarchical propositional structure onto a serial structure, in a way that works despite the limited nature of our short-term memory. Of course, language is not perfect. It is bad at conveying emotion, or details of Euclidian objects: for the latter purpose, a picture is worth a thousand words. But imperfection is no argument against natural selection, as will be obvious to anyone who has suffered from appendicitis or lumbago.

Why is only the capacity to learn a language innate, and nothing more? Would it not be worth while to have the total vocabulary genetically transmitted, and not learnt? One of the possible answers is that, if the vocabulary is learnt, we can acquire names for cultural innovations, such as *screwdriver*. Since cultural evolution is faster than genetic, it is likely that, long before any appreciable number of words could be genetically assimilated, dialects and distinct languages were already present. In Tooby & Cosmides's (1990) phrase, the genome may as well store the vocabulary in the 'cultural environment'.

Piatelli-Palmarini (1989) raised a different objection: since grammar contains many arbitrary elements, it is not a predictable adaptation to communication, and therefore lacks design, and so cannot be a product of natural selection. This misunderstands the nature of evolution by natural selection: in general, the results are not predictable. For example, the gallop of a cheetah, the running of an ostrich and the jumping of a kangaroo are all adaptations to travelling fast on a fairly flat surface. We do not conclude that, because they are different, they are not the products of selection.

Lieberman (1989) argued that universal grammar cannot be the product of natural selection because there is no genetic variation for it. But there is genetic variation. In section 17.8, we describe one example of a qualitative difference that, almost certainly, is genetically caused. Our school memories assure us of quantitative variation in linguistic skill: some people are articulate and others are

not. Evidence for a genetic component in the causation of such quantitative differences is lacking, and would be hard to come by, because children resemble their parents for cultural as well as genetic reasons. But that is no justification for asserting that there is no genetic variance.

It has been argued that language is too sophisticated for its purpose: a simple semantic language would have been sufficient to compete with other species. That is correct, but selection is not mainly interspecific. Humans are social animals. We quoted in section 16.7 Dunbar's evidence that selection for social skills may have been the main motive force for increased brain size in primates, In humans, it would have been selectively advantageous to communicate about time, possession, beliefs, desires, tendencies, obligations, truth, probability, hypotheticals and counterfactuals. The intellectual arms race took place within the species itself. One should also ponder the biological significance of the phrase 'chatting up'.

We now turn to two objections that need to be treated at somewhat greater length. The first is that a 'mutation' for a better syntax would not spread, because at first it would not be understood. A partial answer to this difficulty can be given by pointing to the spread of new words. The word *gay* did not come to mean homosexual because some ruling committee on language decreed that it was so. It spread because at first one person, and then a small group, used it in this sense, and others copied them. We do not react to a new term by complaining that we do not understand, but by trying to find out what it means. European visitors to the USA, confronted by such phrases as *there you go* or *full of shit*, learn their meanings by discovering how they are used and, equally important, by using them experimentally, and judging from the response whether the usage is correct. New words and phrases are memes, able to spread rapidly even if, at first, they are not understood.

But this is only a partial answer to the objection. Changes in grammatical competence that are genetically programmed, not learnt, concern us. The answer to the difficulty perhaps lies in the process of 'genetic assimilation'. As Waddington (1956) pointed out, a trait that first appears as a response to an environmental stimulus may, if it is selectively advantageous, become genetically assimilated, appearing without the stimulus. Experiments have shown that he was right. Essentially, the process works because selection favours those genotypes that respond most readily to the stimulus.

Hinton & Nowlan (1987) simulated genetic assimilation. They imagined a task that initially must be learnt by trial and error, but is gradually converted into one that is genetically programmed, and performed without learning. The population cannot evolve the ability to perform the task purely by natural selection acting on random mutations, without individual learning, because too many mutations must be acquired simultaneously, but evolution is successful if there is a combination of learning and natural selection. As Pinker & Bloom (1990) pointed out, the evolution of language is the ideal field for the operation

of such a process. We do learn new verbal skills, but we also know that, today, many verbal skills are genetically assimilated.

A second difficulty (Chomsky, 1975, and many others) is that grammar must operate as a whole: changes can have drastic consequences. This implies a saltationist origin of grammar: as Bates *et al.* (1989) wrote:

> if the basic structural principles of language cannot be learnt (bottom up) or derived (top down), there are only two possible explanations for their existence. Either universal grammar was endowed to us directly by the Creator, or else our species has undergone a mutation of unprecedented magnitude, a cognitive equivalent of the Big Bang.

This is not an argument likely to appeal to an evolutionary biologist. We have been told too often that the eye could not have evolved by natural selection, because any alteration to it would destroy its function. Yet we know of many intermediates between a simple pigment spot and the vertebrate eye, each one functional. The snag with language is that the intermediates no longer exist. We postpone a discussion of possible intermediates until we have described the gap between animal and human language in more detail. For the moment, we will make only one point. Most of the objections expressed by linguists to possible intermediates amount to saying 'if feature X of grammar was missing, you could not say Y'. This is usually both true and irrelevant. The question is, not whether you could say Y, but whether a grammar without X would be better than no grammar. By analogy, an eye with an iris is better than one without, but an eye without an iris is a lot better than no eye at all.

17.6 Tool use and language: hierarchically organized sequential behaviour

A relationship between tool use and language development was suggested by Greenfield (1991). The first point to establish is that there is a formal similarity between 'action grammar' and protolanguage (Figs 17.4–17.6). The development of action grammar can be illustrated by the way young children manipulate a set of nested cups (Fig. 17.4). Children pass through three stages, of which only the third, the subassembly strategy, is hierarchical. In this strategy, a structure (one cup inside another) that has been produced by one action is treated as a unit in a second action. The analogy between these stages and types of sentence is shown in Fig. 17.5. Again, only the third stage is hierarchical. This is brought out more clearly in Fig. 17.6: the two words, *more grapejuice*, form a single unit, relating to *want*.

The structural analogy is convincing. It is also the case that, both in action grammar and in sentence construction, children acquire these strategies successively, in the same order. However, it is likely that they are carried out by different neural structures, because they do not develop simultaneously: the

Fig. 17.4 Strategies for manipulating nested cups.

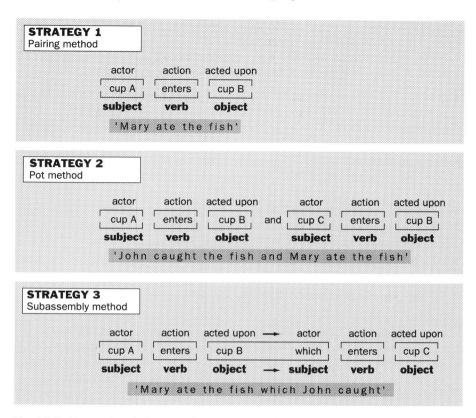

Fig. 17.5 The analogy between action grammar and sentence construction (modified from Greenfield, 1991).

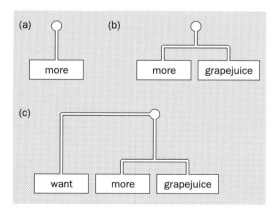

Fig. 17.6 Increasing complexity in early syntax. In (c), *more + grapejuice* form a complex, which is acted on by *want*.

subassembly strategy in the action game appears before the ability to say things such as *want more grapejuice*. Greenfield therefore suggested that the real parallel is between action grammar and word construction.

Children pass through the following stages in word construction:

- The earliest words, towards the end of the first year, are reduplicated consonant–vowel syllables, such as *dada* and *mama*, in which a single consonant is repeatedly combined with the same vowel. At this age, children repeatedly touch one object with another (Piaget, 1952).

- A single consonant is combined with a single vowel, as in *na* (for 'no'). At this age, one object is paired with another, as in the pairing strategy for nesting cups.

- A single consonant is combined successively with two different vowels, as in *baby*. At this stage, children would combine the same cup with two other, different cups, but without ever combining three cups together.

- The initial consonant varies, but the vowel remains constant, as in *kye-bye* ('car bye-bye'). This corresponds to the 'pot' strategy for nesting cups.

- Finally, already developed syllabic subassemblies are combined, as in *ball*.

Further evidence for a connection between the two processes comes from aphasias (speech defects) resulting from damage to Broca's area, located in the left frontal lobe of the cerebral cortex. Adults who have suffered damage to this area are often incapable of producing syntactically organized speech: their utterances lack any trace of a tree structure. They often show a similar failure in object manipulation: an example is shown in Fig. 17.7. It is relevant to our earlier discussion of the parallel between action grammar and word formation that Lieberman (1984) argued that Broca's area is concerned not only with syntax but also with phonology. Although an inability to deal with hierarchies is associated with damage to Broca's area, and not with other types of damage, there are complications. The association between object and sentence manipulation is stronger in children than in adults: some adults are good at syntax and bad at action grammar, and vice versa. Greenfield suggests that early in development both functions are performed by Broca's area, but later the functions are separated, and taken over by an area in the prefrontal region.

How do apes perform on this 'pairing–pot–subassembly' scale? Although many primates use tools, only chimpanzees use the same tool on different objects, and two tools sequentially on the same object. This 'pot' strategy is also apparent when they use a stone to crack a nut placed on an anvil (Sugiyama & Koman, 1979). The technique is taught by mothers to their offspring, and teaching is accompanied by gestural communication (Boesch, 1993)—the only known example of explicit animal pedagogy in the wild.

There is no evidence that chimpanzees use the subassembly strategy in the wild, but they occasionally do so in captivity. The first record is Koehler's (1925)

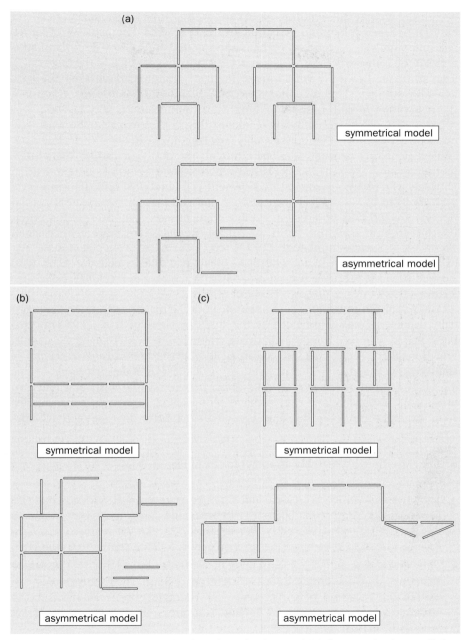

Fig. 17.7 Diagrams drawn by (b) Broca's aphasics, and (c) fluent aphasics, suffering damage to Wernicke's area. In both cases, the diagrams below were drawn when trying to reproduce the models in (a).

account of a chimpanzee fitting two sticks together in order to make a tool long enough to reach a banana. Matsuzawa (1991) studied chimpanzees, of varied age and linguistic experience, manipulating nested cups. They discovered strategies in the same order that children do, but usually got stuck at the 'pot' stage. Two of them, however, with intense linguistic training, discovered the subassembly strategy. This hint that there may be mutual reinforcement between the acquisition of skills in manipulating words and objects is very exciting.

In using tools, then, chimpanzees adopt the 'pot' strategy in the wild, and occasionally achieve the subassembly strategy in captivity. Their linguistic achievements are probably similar, although the data—or rather, their interpretation—are disputed. Plooij (1978) reported some gestural communication between wild chimps involving two-element combinations, and the same level was observed by Boesch between mother and child. Such two-element combinations are certainly observed in captive chimpanzees. Sentences with a primitive hierarchical structure, as in *want more grapejuice*, are more controversial, Greenfield & Savage-Rumbaugh (1991) report examples of such sentences. For example, a chimpanzee that had been playing at hiding balloons filled with Koolaid used the sentence *balloon water hide*, in which *water* modifies *balloon*, to form a subassembly acting as the object of *hide*.

An idea of the maximum command of syntax achieved by captive, linguistically trained animals is provided by the study by Herman *et al.* (1984) of two bottle-nosed dolphins: one, Phoenix, was trained to respond to acoustic signals, and the other, Areakamai, to gestures. They performed only as receivers of signals, not as senders. Phoenix had a vocabulary of some 30 words (Table 17.1). Note that she learnt to distinguish between verbs with one and two arguments. Both dolphins responded to sentences of from two to five words: some examples are given in Table 17.2. To summarize their syntactic competence:

- They use the subassembly strategy; for example, in treating *surface pipe* as a unit.

- They use word order to indicate meaning: for example, *hoop fetch pipe* and *pipe fetch hoop* are interpreted differently. Phoenix and Areakamai were taught to interpret word order differently. For example, Areakamai required *pipe hoop fetch* (and not *hoop fetch pipe*) as an instruction to fetch the hoop to the pipe.

- They must discover the meaning of a sentence by analysing the whole structure, and not simply by left-to-right processing. Thus, later words can modify the meaning of earlier ones: for example, for Areakamai, in the sentence *basket over*, the word 'basket' is a direct object, whereas in *basket hoop fetch*, 'basket' is a goal.

- The dolphins were good at comprehending novel sentences and syntactic forms. For example, Phoenix had been trained to understand the syntactic

Table 17.1 The vocabulary of Phoenix, a bottle-nosed dolphin

Objects

1. Tank fixtures: *gate*, *panel*

2. Objects that can be moved by the trainers: *speaker*, *water* (jetted from hose), *net*, *Areakamai* (the other dolphin)

3. Objects that Phoenix can move: *ball*, *hoop*, *pipe*, *fish*, *frisbee*, *surfboard*, *basket*, *person*

Actions

1. Taking a direct object only: *tail-touch*, *pectoral-touch*, *mouth* (grasp with mouth), *(go) under*, *(go) over*, *(go) through*, *toss*, *spit*

2. Taking direct and indirect objects: *fetch*, *in* (place one object in or on another)

Modifiers

right or *left*, *surface* or *bottom* (used before object to indicate its location)

Agents

Phoenix, *Areakamai* (prefix for each sentence; calls dolphin to station; indicates which dolphin is to receive fish reward)

Other

erase (cancels preceeding words)

yes (after correctly executed instruction)

no (sometimes used after incorrectly executed instruction: can cause emotional behaviour)

After Herman *et al.*, 1984.

Table 17.2 Sentences understood by Phoenix

Rule	Examples (meaning)
Two-word	
Direct object + action (DO + A)	*basket tail-touch* (touch basket with tail-fluke)
Three-word	
DO + A + Indirect object (DO + A + IO)	*hoop fetch pipe** (fetch the hoop to the pipe)
Four-word	
Modifier + DO + A + IO (M + DO + A + IO)	*bottom hoop fetch panel** (fetch the hoop on the bottom to the panel)
Five-word	
(M + DO + A + M + IO)	*surface pipe fetch bottom hoop** (fetch the pipe on the surface to the hoop on the bottom)

* Reversal of order of direct and indirect objects reverses the meaning.
After Herman *et al.*, 1984.

form Modifier + Direct object + *fetch* + Indirect object. Later, she was given the form Direct object + *fetch* + Modifier + Indirect object, without specific training. She responded correctly to the first sentence of this type.

- There is preliminary evidence that dolphins can understand conjoined sentences, such as *Phoenix pipe tail-touch pipe over*, as 'first touch the pipe with the tail and then jump over it'. It is interesting that sometimes Phoenix performed the two actions in reverse order.

In the face of these observations it cannot be claimed that animals have no syntactic competence. However, the performance was in receiving signals, not in sending them, and it was achieved by linguistically trained captive animals. We have little idea of how dolphins communicate in the wild.

17.7 Brain damage and language disorders

We have already mentioned the loss of grammatical competence resulting from damage to Broca's area. Very different are the 'fluent aphasics' resulting from damage to Wernicke's area. These patients utter a meaningless but syntactically correct babble. Figure 17.7b shows that their manipulation of objects suffers from a similar inability to remember content, despite a correct structure. Other examples are known in which damage to a particular region of the brain is associated with a specific impairment of linguistic competence (for a recent review, see Damasio & Damasio, 1992). In this section we will confine ourselves to one particular type of deficiency. The ability to name concepts requires three abilities: to form concepts, to form words, and to form appropriate connections between words and concepts. We will discuss only aphasias in which the link between word and concept is broken.

Patients with damage in the temporal segment of the left lingual gyrus suffer from 'colour anomia'. They experience colour normally, and are in command of word morphology, but are unable to pair names with colours: for example, they may pair yellow with grass and green with banana. Given a colour name, the patient will point to the wrong colour. The link between word and concept is impaired.

Patients with damage to the anterior and mid-temporal cortices recognize objects well, but cannot name them: they may say, for example, 'I know it, but I cannot say the name'. Oddly enough, this inability to name things is more pronounced for natural than for human artefacts, suggesting that the brain may use different parts to represent objects belonging to these categories.

Damage to the left anterior temporal lobe may impair the ability to remember the names of unique persons.

Such evidence for an association between particular linguistic abilities and particular regions of the brain reinforces the view that linguistic competence is

specific, and not merely an aspect of general cognitive ability. We now turn to some genetic evidence, which has an even more direct bearing on the nature of language and its evolution.

17.8 The genetics of language disorders

Gopnik (1990; see also Gopnik & Crago, 1991) described an English-speaking family among whom a peculiar language disorder (dysphasia) occurs, in 16 family members out of 30 over three generations (Fig. 17.8). The data are consistent with the view that the disorder is caused by a single dominant autosomal gene. The genetic interpretation is strengthened by the fact that some members of a sibship may be affected while others are normal: this rules out the idea that the trait is familial because children learn from their parents. Dysphasics remain disordered even if they receive special linguistic training, and even if one parent is normal. There is no indication that other cognitive abilities are affected: these people tell jokes and converse, and may even be mathematically able. There is no general failure to handle hierarchical structures.

Gopnik has used a range of tests to diagnose the condition, but its nature is best explained by quoting some sentences written by affected children (we have slightly shortened some of these, we hope without altering their significance):

> She remembered when she hurts herself the other day.
> Carol is cry in the church.
> On Saturday I went to nanny house with nanny and Carol.

In each of these sentences the child has failed to make an appropriate change in the form of a word: in the first two, a change is required to indicate the past tense (hurt, crying), and in the third to express ownership (nanny's). Affected children have the same difficulty with plurals. A child may learn that a picture of a single book is a 'book', and of several books are 'books'. The child is then shown a picture of an imaginary animal, and is told that it is a 'wug': if then shown a picture of several wugs, the child does not know that the appropriate word is 'wugs'. Thus the child can learn particular examples of singular and plural, or of tense, just as we all have to learn the meanings of particular lexical items such as 'horse' and 'cow', but does not generalize.

The failure to generalize is nicely illustrated by the following anecdote. Writing an account of what she did at the weekend, a child wrote:

> On Saturday I watch TV.

Admittedly, this could be seen as a grammatically correct statement about what she is accustomed to do on Saturdays. Reasonably, however, the teacher treated it as a statement about what she had done the previous weekend, and

The *Origin of Language*

Fig. 17.8 Pedigree of a family with a specific language impairment. Affected individuals are shown by filled circles. (Data from Gopnik, 1990.)

corrected it to 'watched'. Next weekend, the child wrote:

On Saturday I wash myself and I watched TV and I went to bed.

Three points emerge. She has learnt that the past tense of watch is 'watched'; she has failed to generalize to 'washed'; and she already knows that the past of go is 'went': after all, this is something we must all learn as a unique fact, not by generalization.

This fascinating case has some important implications. First, although affected people are impaired, they are not without grammar: they are a great deal better off than they would be if they could not talk at all. In other words, there can be intermediates between perfect competence and no competence. Second, the impairment is specific to language: there is no mental defect. This confirms Chomsky's view that linguistic competence is not a mere spin-off of general intelligence. Third, it suggests a road to understanding the evolution of language. But we badly need other comparable gene loci: unfortunately, if they do not show dominant inheritance, they will be a lot harder to find.

17.9 Protolanguage

Language does not fossilize. There are, nevertheless, four sources that may indicate what primitive language was like: ape languages, early child language, language of 'Caspar Hauser' children and pidgin languages. We discusses them in turn.

Consider the following utterances, from two different sources:

<div align="center">

Big train; Red book. (12)
Adam checker; Mommy lunch.
Walk street; Go store.
Adam put; Eve read.
Put book; Hit ball.

Drink red; Comb black. (13)
Clothes Mrs G; You hat.
Go in; Look out.
Roger ticket; You drink.
Tickle Washoe; Open blanket.

</div>

Although remarkably similar, (12) is from children at the two-word stage, and (13) is from the chimpanzee Washoe (Gardner & Gardner, 1974). They illustrate the following types of utterance: the attribution of qualities to objects; the possession of non-living objects by living beings; the location of actions; the relation of agents and patients to actions. The two sets are formally identical.

In both apes and in children under two, one-word utterances dominate. There are no grammatical items and no trace of structure. The only real difference

between (12) and (13) is that children seem to categorize for categorization's sake, whereas apes talk only about objects they want or actions they want to perform or have performed: children have a greater curiosity about the environment.

There have been many examples of children growing up in isolation, but the best-documented case linguistically is that of 'Genie', a 13-year-old girl who was found in California in 1970. Her father imprisoned her in her room alone when she was about 18 months old. She was incapable of speech. Although normal in her ability to form concepts, she remained stuck at the linguistic level illustrated below:

> Want milk, (14)
> Mike paint.
> Appplesauce buy store.
> At school wash face.
> Very sad, climb mountain.
> I want Curtiss play piano.

Genie is a little more advanced than (12) or (13): there is one grammatical item, *at*, and the last sentence is a fairly complex proposition. But the difference is slight: it seems that Genie has acquired something less than human language, which we can call protolanguage. In view of the similarity to ape language, protolanguage is phylogenetically ancient. There is apparently no critical period at which it must be acquired.

Turning to our final example, pidgins develop when people must communicate with one another without sharing a common language, and without one language being the obvious medium. Slaves in island colonies are a good example. The following comes from Bickerton (1990):

> *Luna*, hu *hapai*? *Hapai* awl, hemo awl.
> Foreman, who carry? Carry all, cut all.

('Who will carry it, boss? Everyone will cut it, and everyone will carry it.') The speaker of the first line was a Filipino, while the italicized words are Hawaiian. Although there are a few grammatical items such as *hu*, there are no articles, prepositions or tense markers.

Sometimes pidgin remains at a primitive level for several generations. Russonorsk, of which an example follows, was used by Russian and Scandinavian sailors for trade:

R: What say? Me no understand.
N: Expensive, Russian—goodbye.
R: Nothing. Four half.
N: Give four, nothing good.
R: No brother. How me sell cheap? Big expensive flour on Russia this year.

N: You no say true.

R: Yes. Big true, me no lie, expensive flour.

N: If you buy—please four pud. If you no buy—then goodbye.

R: No, nothing brother, please throw on deck.

There is some advance here over (12) and (13), but there are few grammatical items, the longest utterance has no verb, *give* lacks a goal, and *throw* a patient. The remarkable thing about pidgin is how little linguistic knowledge its speakers were able to employ when communicating across language barriers.

The crucial differences between the various forms of protolanguage and true language are as follows:

- Word order in protolanguage has nothing to do with syntax.

- In ordinary language, null elements indicate the points in sentences where we can infer that some constituent is notionally present. In protolanguage, any element may be missing from any position in an unpredictable way.

- Phrase structure is almost completely absent. The few apparent exceptions may have been rote-learned, such as lexical items and idioms.

- There are few, if any, grammatical items.

17.10 From protolanguage to language

Before discussing the evolutionary transition from protolanguage to language, we will describe a comparable transition that can occur in a single generation. This is the transition from pidgin to creole (Bickerton, 1983). The latter is a transformed pidgin, created by children when learning pidgin from adults. Children from Hawaiian pidgin-speaking parents utter sentences such as:

$$\text{They wen go up there early in the morning—go plant} \qquad (15)$$

('They went up there early in the morning to plant (crops).') The null element refers to the missing subject, *they*.

Creole languages arose when children of immigrants were not exposed to any normal natural language, but to pidgin only. Table 17.3 gives some examples of the jump in complexity between pidgin and Hawaiian Creole English. Whereas almost everyone speaks his or her own pidgin, the rules of Creole are uniform from speaker to speaker. One might expect that the grammar comes either overwhelmingly from one source language, or that it is a mixture derived from all the available source languages, but that is not so. Hawaiian Creole differs from Chinese, Hawaiian, Korean, Portugese, Spanish, or the Philippine languages: some differences between English and Hawaiian Creole are shown in Table 17.4.

Creole languages have arisen repeatedly in various parts of the world, with widely different vocabularies, but with a surprisingly similar grammatical

Table 17.3 Some sentences with identical meanings in pidgin and Hawaiian Creole English

Pidgin	Hawaiian Creole English
Building—high place—wall part—time—now time—and then—now temperature every time give you	Get one electric sign high up on da wall of da building show you what time an temperature get right now
Now days ah house ah inside wash clothes machine get, no? Before time ah no more, see? and then pipe no more water pipe no more	Those days bin get no more washing machine pipe water like get inside house nowadays ah?

From Bickerton, 1983.

Table 17.4 Grammatical differences between sentences in English and Hawaiian Creole English

English	Hawaiian Creole English
The two of us had a hard time raising dogs	Us two bin get hard time raising dogs
John and his friends are stealing the food	John-them stay cockroach the kau-kau
How do you expect to finish your house?	How you expect for make pay you house?
There was a woman who had three daughters	Bin get one wahine she get three daughter
She can't go because she hasn't any money	She no can go, she no more money, a's why

From Bickerton, 1983.

structure. One characteristic feature of Creoles (and, as it happens, Hungarian) is that they permit the double negative, 'I don't want no cabbage'. A second common feature (also shared by Hungarian) is to distinguish statements and questions by intonation alone. A third grammatical feature shared by many Creoles is the method of conjugating verbs (Table 17.5). Bickerton (1983) has drawn attention to the fact that, between the ages of 2 and 4, children learning English sometimes make mistakes that would be grammatical in Creole: for example, children often try out the double negative.

Creoles are proper languages: for example, the proportion of purely grammatical items is about 50 per cent. Apparently, Hawaiians jumped from protolanguage to language in a single generation. This rapid transition was possible because no genetic change was needed: the change could not have been so rapid in phylogeny.

Table 17.5 Common features in the conjugation of verbs in Haitian Creole and Sranan (an English-based Creole spoken in Surinam); Hawaiian Creole is similar

English meaning	Haitian Creole	Sranan
He walked	Li mache	A waka
He had walked	Li te mache	A ben waka
He will/would walk	L'av(a) mache	A sa waka
He is/was walking	L'ap mache	A e waka
He would have been walking	Li t'av ap mache	A ben sa e waka

After Bickerton, 1983.
Note the similar order of ben = te, sa = av, e = ap, in the construction of 'he would have been walking': Bickerton points out other similarities in the conjugation of verbs.

Given the rather sharp distinction between protolanguage and language, we must return to the question whether any functional intermediates between them are possible. We are tempted to point to Gopnik's dysphasics, who lack a specific grammatical skill but are obviously better off than speakers like Genie, and leave it at that. However, the difficulty is so often referred to by linguists that we will say a little more about it.

The difficulty arises, we believe, because people are asking the wrong question. They ask: is there any grammatical rule that could be absent, without reducing the power of linguistic expression? The answer is probably no, but the question is irrelevant. We should ask: is there any grammatical rule that, if it were absent, would still leave a linguistic competence that was better than mere protolanguage? To this question, the answer is obviously yes.

To give one example, Premack (1985) suggested a language in which thematic roles would map directly onto surface ordering. It would then be possible to say *the dog bit John,* but not *John was bitten by the dog.* Such a grammar would be less expressive than our existing grammar, but it would be better than no grammar. Language would be still simpler if it permitted only verbs with two arguments, agent and patient. One could then say *the dog bit John,* but not *John sleeps,* or *John gave the book to Mary*: again, the limitations are largely irrelevant.

Consider the following possible extensions to protolanguage:

- Items for the negation of information, such as *no.*
- 'Wh–' questions, such as *what, who* and *where.*
- Pronouns (instead of the repetition of proper names: *John asked Bill for the book and Bill gave it to him,* instead of *John asked Bill for the book and Bill gave it to John*).
- Verbal auxiliaries such as *can* and *must.*

- Expressions for earlier and later (e.g. *before* and *after*).
- Expressions for orientation in space (e.g. *up*, *down* and *into*).
- Quantities such as *many* and *few*.

There seems no good reason why these features should not have been acquired one by one. The hard question, however, is how we came to use a three-tiered phrase structure (Figs 17.2 and 17.3), and to construct sentences by recursively inserting phrases into phrases. The crucial invention was the template for phrase assembly. The three layers (Fig. 17.3) represent, respectively, a generic class (*cow*), properties peculiar to particular members of that class (*with a crumpled horn*) and specification of the complete individual. Thus, phrases are machines for going from the class to the individual, an ability important for cognition as well as for communication. The template may have originated with noun phrases, and later been extended to other units, so that one could say *the cow with the crumpled horn kicked the dog*, but not *the dog ran round the mulberry bush*, which contains a verb phrase.

It is probably foolish to attempt a reconstruction of the precise path that was followed from protolanguage to language, at least until we have more extensive genetic data. Our only aim has been to show that it is not particularly difficult, in principle, to imagine a stepwise transition.

17.11 Conclusions

Evolution from apes to humans has involved both an increase in general cognitive ability, and in specific competence for language. The steady increase in brain size over the past 2 million years suggests that the increase in cognitive ability also proceeded over this period. The main selective advantages conferred were improved ability in the manufacture and use of tools, and in social skills. Already, in other primates, there is evidence of an association between brain size and the complexity of the social system, and this association was almost certainly important in our own ancestors.

The timing of the origin of language is more difficult to determine. However, the dramatic increase in technical inventiveness during the past 40 000 years, discussed in the last chapter, is most readily explained if the final stages in linguistic competence emerged at that time. Humans differ from apes not only in grammatical competence, which has been the main topic of this chapter, but also in the ability to produce and perceive sounds. One difficulty in teaching apes to talk arises because they cannot make all the sounds used in human speech. Human evolution required both anatomical changes, and improvements in the brain mechanisms concerned with sound production and perception (Lieberman, 1989). Anatomically, the descent of the larynx in humans has increased the range of sounds we can make, at the cost of increasing the chance

of choking when we eat or drink. It would be helpful if we could date this change from the fossil record, but anatomists are divided as to how far this is possible: unfortunately, the hyoid cartilage, which would provide the information, rarely fossilizes.

The changes in brain mechanisms are equally important. Although sound input is in principle a continuous variable, we unconsciously classify incoming speech sounds into discrete categories: in other words, we treat the input as digital. We are able to produce and to perceive sound 'segments' (roughly equivalent to the letters of the alphabet) at the astonishingly high rate of 25 per second. This rapid transmission is necessary: without it, we would forget the beginning of a sentence before reaching the end.

The emergence of human language, then, required changes in anatomy, in motor control, in sound perception, and in grammatical competence. These changes could not have been instantaneous, but may well have been rapid. They must have predated the dispersal of *Homo sapiens* throughout the world, most probably from Africa, because all existing human populations are alike in these respects. Subject to this constraint, however, the events may have been fairly recent.

This book has been about the major changes that have taken place in the way in which information is encoded, and transmitted between generations. The origin of language is the last transition of this kind that had a genetic basis, but it is not the last transition. The invention of writing made possible the emergence of modern, large-scale societies, and the change from societies dominated by magic and ritual to those in which science and reason play an increasing role (Goody, 1977). We are today in the midst of yet another major transition, to a society in which information is stored and transmitted electronically. It is impossible to foresee where this latest transition will lead. But the emergence of computer viruses may be a straw in the wind: we must beware that we are not replaced by a new kind of self-replicating entity.

References

Adams, J. & Hansche, P. E. 1974. Population in microorganisms. I. Evolution of diploidy in *Saccharomyces cerevisiae*. *Genetics* 76: 327–38.

Akam, M. 1989a. Making stripes inelegantly. *Nature* 341: 282–3.

Akam, M. 1989b. *Hox* and *HOM*: homologous gene clusters in insects and vertebrates. *Cell* 57: 347–9.

Alexander, R. D. 1974. The evolution of social behavior. *Annual Review of Ecology and Systematics* 5: 325–83.

Anxolabehere, D., Kidwell, M. G. & Periquer, G. 1988. Molecular characteristics of diverse populations are consistent with the hypothesis of a recent invasion of *Drosophila melanogaster* by mobile P elements. *Molecular Biology and Evolution* 5: 252–69.

Aoun, J. 1992. A brief representation of the generative enterprise. In: J. A. Hawkins & M. Gell-Mann (eds), *The Evolution of Human Languages*, SFI Studies in the Sciences of Complexity, Proc. Vol. X, pp. 121–35. Addison-Wesley, Reading, Mass.

Arendt, D. & Nübler-Jung, K. 1994. Inversion of dorso-ventral axis? *Nature* 371: 26.

Axelrod, R. 1984. *The Evolution of Cooperation*. Basic Books, New York.

Axelrod, R. & Hamilton, W. D. 1981. The evolution of cooperation. *Science* 211: 1390–6.

Bachmann, P. A., Luisi, P. L. & Lang, J. 1992. Autocatalystic self-replicating micelles as models for prebiotic structures. *Nature* 357: 57–9.

Baker, A. & Schatz, G. 1987. Sequences from a prokaryotic genome or the mouse dehydrofolate reductase gene can restore the import of a truncated precursor protein into yeast mitochondria. *Proceedings of the National Academy of Sciences USA* 84: 3117–21.

Bangham, A. D. & Horne, R. W. 1964. Negative staining of phospholipids and their structural modification by surface-active agents as observed in the electron microscope. *Journal of Molecular Biology* 8: 660–8.

Bartz, S. H. & Hölldobler, B. 1982. Colony founding in *Myrmecocystus mimicus* Wheeler (Hymenoptera, Formicidae) and the evolution of foundress associations. *Behavioural Ecology and Sociobiology* 10: 137–47.

Bates, E., Thal, D. & Marchman, V. 1989. Symbols and syntax: a Darwinian approach to language development. In: N. Krasnegor, D. Rumbaugh, M. Studdert-Kennedy & R. Schiefelbusch (eds), *The Biological Foundations of Language Development*, pp. 29–65. Oxford University Press, Oxford.

Bateson, W. 1984. *Materials for the Study of Variation*. Macmillan, London.

Beardsley, T. 1991. Smart genes. *Scientific American* 265(2): 86–95.

Beeby, R. & Kacser, H. 1990. Metabolic constraints in evolution. In: J. Maynard Smith &

G. Vida (eds), *Organizational Constraints on the Dynamics of Evolution*, pp. 57–75. Manchester University Press, Manchester.

Bell, G. 1982. *The Masterpiece of Nature*. Croom Helm, London.

Bell, G. 1985. The origin and early evolution of germ cells as illustrated by the Volvocales. In: H. Halvorson & A. Mornoy (eds), *The Origin and Evolution of Sex*, pp. 221–56. Alan R. Liss, New York.

Bell, G. & Burt, A. 1990. B-chromosomes: germ-line parasites which induce changes in host recombination. *Parasitology* 100: S19–S26.

Bell, G. & Koufopanou, V. 1991. The architecture of the life cycle in small organisms. *Philosophical Transactions of the Royal Society of London* B332: 81–9.

Bell, G. & Maynard Smith, J. 1987. Short-term selection for recombination among antagonistic species. *Nature* 328: 66–8.

Bengtsson, B. O. 1985. Biased conversion as the primary function of recombination. *Genetical Research, Cambridge* 47: 77–80.

Benner, S. A., Allemann, R. K., Ellington, A. D., Ge, L., Glasfeld, A., Leanz, G. F. *et al.* 1987. Natural selection, protein engineering, and the last ribo-organism: rational model building in biochemistry. *Cold Spring Harbor Symposia on Quantitative Biology* 52: 56–63.

Benner, S. A., Ellington, A. D. & Tauer, A. 1989. Modern metabolism as a palimpsest of the RNA world. *Proceedings of the National Academy of Sciences, USA* 86: 7054–8.

Bernstein, H., Byers, G. S. & Michod, R. E. 1981. Evolution of sexual reproduction: importance of DNA repair, complementation, and variation. *American Naturalist* 117: 537–49.

Bernstein, H., Hopf, F. A. & Michod, R. E. 1988. Is meiotic recombination an adaptation for repairing DNA, producing genetic variation, or both? In: R. E. Michod & B. R. Levin (eds), *Evolution of Sex: An Examination of Current Ideas*, pp. 106–25. Sinauer, Sunderland, Mass.

Bickerton, D. 1983. Creole languages. *Scientific American* 249(1): 108–15.

Bickerton, D. 1990. *Language and Species*. University of Chicago Press, Chicago.

Biebricher, C. K., Eigen, M. & Luce, R. 1981. Product analysis of RNA generated *de novo* by Q-beta replicase. *Journal of Molecular Biology* 148: 369–90.

Biebricher, C. K., Eigen, M. & Gardiner, W. C. Jr. 1985. Kinetics of RNA replication: competition and selection among self-replicating RNA species. *Biochemistry* 24: 6550–60.

Blake, C. C. F. 1985. Introns and the evolution of proteins. *International Review of Cytology* 93: 149–85.

Blobel, G. 1980. Intracellular membrane topogenesis. *Proceedings of the National Academy of Sciences, USA* 77: 1496.

Boerlijst, M. C. & Hogeweg, P. 1991. Spiral wave structure in pre-biotic evolution— hypercycles stable against parasites. *Physica* D48: 17–28.

Boesch, C. 1993. Aspects of transmission of tool-use in wild chimpanzees. In: K. R. Gibson & T. Ingold (eds), *Tools, Language and Cognition in Human Evolution*, pp. 171–83. Cambridge University Press, Cambridge.

Bonner, J. T. 1982. Evolutionary strategies and developmental constraints in the cellular slime molds. *American Naturalist* 119: 530–52.

Boyd, R. & Richerson, P. J. 1985. *Culture and the Evolutionary Process*. University of Chicago Press, Chicago.

Bresch, C., Niesert, V. & Harnasch, D. 1980. Hypercycles, parasites and packages. *Journal of Theoretical Biology* 85: 399–405.

Briggs, D. & Walters, S. M. 1984. *Plant Variation and Evolution*. Cambridge University Press, Cambridge.

Broek, D., Bartlett, R., Crawford, K. & Nurse, P. 1991. Involvement of p34cdc2 in establishing the dependency of S phase on mitosis. *Nature* 349: 388–93.

Brookfield, J. F. Y. 1991. Models of repression of transposition in P-M hybrid disgenesis by P cytotype and by zygotically encoded repressor proteins. *Genetics* 128: 471–86.

Bryant, P. J. 1993. The polar coordinate model goes molecular. *Science* 259: 471–2.

Bryant, S. V., French, V. & Bryant, P. J. 1981. Distal regeneration and symmetry. *Science* 212: 993–1002.

Buckle, G. R. & Greenberg, L. 1981. Nestmate recognition in sweat bees (*Lasioglossum zephyrum*): does an individual recognize its own odour or only odour of its nestmates? *Animal Behaviour* 29: 802–9.

Bull, J. J., Molineux, I. L. & Rice, W. R. 1991. Selection of benevolence in a host–parasite system. *Evolution* 45: 875–82.

Bulnheim, H. P. 1978. Interaction between genetic, external, and parasitic factors in sex determination of the crustacean amphipod *Gammarus duebeni*. *Helgölander wissenschaftlich Meerersuntersuchungen* 31: 1–33.

Buss, L. W. 1987. *The Evolution of Individuality*. Princeton University Press, Princeton.

Butlerow, A. 1861. Formation synthétique d'une substance sucrée. *Comptes rendus de l'Académie des Sciences, Paris* 53: 145–7.

Cairns-Smith, A. G. 1971. *The Life Puzzle*. Oliver & Boyd, Edinburgh.

Campbell, A. M. 1993. Genome organization in prokaryotes. *Current Opinion in Genetics and Development* 3: 837–44.

Cann, R. L., Stoneking, M. & Wilson, A. C. 1987. Mitochondrial DNA and human evolution. *Nature* 325: 31–6.

Canning, E. U. 1988. Nuclear division and chromosome cycles in microsporidia. *Biosystems* 21: 333–40.

Carpenter, A. T. C. 1987. Gene conversion, recombination nodules, and the initiation of meiotic synapsis. *BioEssays* 6: 232–6.

Castets, V., Dulos, E., Boissonade, J. & De Kepper, E. 1990. Experimental evidence of a sustained standing Turing-type nonequilibrium chemical pattern. *Physical Review Letters* 64: 2953–6.

Cavalier-Smith, T. 1985. *The Evolution of Genome Size*. Wiley, Chichester.

Cavalier-Smith, T. 1987a. The origin of eukaryotic and archaebacterial cells. *Annals of the New York Academy of Sciences* 503: 17–54.

Cavalier-Smith, T. 1987b. Bacterial DNA segregation: its motors and positive control.

Journal of Theoretical Biology 127: 361–72.

Cavalier-Smith, T. 1987c. The origin of cells; a symbiosis between genes, catalysts, and membranes. *Cold Spring Harbor Symposia on Quantitative Biology* 52: 805–42.

Cavalier-Smith, T. 1987d. The simultaneous symbiotic origin of mitochondria, chloroplasts, and microbodies. *Annals of the New York Academy of Science* 503: 55–71.

Cavalier-Smith, T. 1988. Eukaryote cell evolution. *Proceedings 13th International Botanical Congress*, pp. 203–23.

Cavalier-Smith, T. 1991a. The evolution of cells. In: S. Osawa & T. Honjo (eds), *Evolution of Life: Fossils, Molecules and Culture*, pp. 271–304. Springer-Verlag, Tokyo.

Cavalier-Smith, T. 1991b. Intron phylogeny: a new hypothesis. *Trends in Genetics* 7: 145–8.

Charlesworth, B. & Hartl, D. L. 1978. Population dynamics of the segregation distorter polymorphism of *Drosophila melanogaster*. *Genetics* 89: 171–92.

Charlesworth, B. & Langley, C. H. 1989. The population genetics of *Drosophila* transposable elements. *Annual Review of Genetics* 23: 251–87.

Chomsky, N. 1957. *Syntactic Structures*. Mouton, The Hague.

Chomsky, N. 1968. *Language and Mind*. Harcourt, Brace and World, New York.

Chomsky, N. 1975. *Reflections on Language*. Pantheon Books, New York.

Chyba, C. F., Thomas, P. J., Brookshaw, L. & Sagan, C. 1990. Cometary delivery of organic molecules to the early Earth. *Science* 249: 366–73.

Clausen, J. & Hiesey, W. M. 1958. Experimental studies on the nature of species. IV. Genetic structure of ecological races. *Carnegie Institution of Washington Publication* No. 615.

Clay, K. 1988. Clavipitaceous fungal epiphytes of grasses: coevolution and the change from parasitism to mutualism. In: Pirozynski & Hawksworth (op. cit.), pp. 79–105.

Cleveland, L. R. 1947. The origin and evolution of meiosis. *Science* 105: 287–9.

Coen, E. S. & Meyerowitz, E. M. 1991. The war of the whorls: genetic interactions controlling flower development. *Nature* 353: 31–7.

Cohan, F. M., Roberts, M. S. & King, E. C. 1991. The potential for genetic exchange by transformation within a natural population of *Bacillus subtilis*. *Evolution* 45: 1393–421.

Conway Morris, S. 1993. The fossil record and the early evolution of the metazoa. *Nature* 361: 219–25.

Cosmides, L. M. & Tooby, J. 1981. Cytoplasmic inheritance and intragenomic conflict. *Journal of Theoretical Biology* 89: 83–129.

Couvet, D., Bonnemaison, F. & Gouyon, P. H. 1986. The maintenance of females among hermaphrodites: the importance of nuclear–cytoplasmic interactions. *Heredity* 57: 325–30.

Coyne, J. A. & Orr, H. A. 1989. Patterns of speciation in *Drosophila*. *Evolution* 43: 362–81.

Crain, S. 1991. Language acquisition in the absence of experience. *Behavioral Brain Science* 14: 597–650.

Crick, F. H. C. 1968. The origin of the genetic code. *Journal of Molecular Biology* 38: 367–79.

Crow, J. F. 1991. Why is Mendelian segregation so exact? *BioEssays* 13: 305–12.

Cullis, C. A. 1983. Variable DNA. In: K. F. Chater *et al.* (eds), *Genetic Rearrangements*, pp. 253–64. Croom Helm, London.

Damasio, A. R. & Damasio, H. 1992. Brain and language. *Scientific American* 267(3): 89–95.

Darlington, C. D. 1939. *Evolution of Genetic Systems.* Cambridge University Press, Cambridge.

Darnell, J. E. & Doolittle, W. F. 1986. Speculations on the early course of evolution. *Proceedings of the National Academy of Sciences, USA* 83: 1271–5.

Darnell, J. E., Lodish, H. & Baltimore, D. 1986. *Molecular Cell Biology.* W. H. Freeman, New York.

Darwin, C. 1877. *The Different Forms of Flowers on Plants of the Same Species.* John Murray, London.

Davidson, E. H. 1990. How embryos work: a comparative view of diverse modes of cell fate specification. *Development* 108: 365–89.

Dawkins, R. 1976. *The Selfish Gene.* Oxford University Press, Oxford.

Derry, L. A., Kaufman, A. J. & Jacobsen, S. B. 1992. Sedimentary cycling and environmental change in the late Proterozoic—evidence from stable and radiogenic isotopes. *Geochimica et Cosmochimica Acta* 56: 1317–29.

Di Giulio, M. 1989. The extension reached by the minimization of the polarity distances during the evolution of the genetic code. *Journal of Molecular Evolution* 29: 288–93.

Dobzhansky, Th. 1951. *Genetics and the Origin of Species*, 3rd edn. Columbia University Press, New York.

Domingo, E., Davila, M. & Ortin, J. 1980. Nucleotide sequence heterogeneity of the RNA from a natural population of foot-and-mouth-disease virus. *Gene* 11: 333–46.

Doolittle, W. F. & Sapienza, C. 1980. Selfish genes, the phenotype paradigm and genome evolution. *Nature* 284: 601–3.

Doudna, J. A. & Szostak, J. W. 1989. RNA-catalysed synthesis of complementary-strand RNA. *Nature* 339: 519–52.

Doudna, J. A., Couture, S. & Szostak, J. W. 1991. A multisubunit ribozyme that is a catalyst of and template for complementary strand RNA synthesis. *Science* 251: 1605–8.

Douglas, A. E. & Smith, D. C. 1989. Are endosymbioses mutualistic? *Trends in Ecology and Evolution* 4: 350–2.

Douglas, S. E., Murphy, C. A., Spencer, D. F. & Gray, M. W. 1991. Cryptomonad algae are evolutionary chimaeras of two phylogenetically distinct unicellular eukaryotes. *Nature* 350: 148–51.

Drake, J. W. 1991. A constant rate of spontaneous mutation in DNA-based microbes. *Proceedings of the National Academy of Sciences, USA* 88: 7160–4.

Driever, W. & Nüsslein-Volhard, C. 1988. The bicoid protein determines position in the *Drosophila* embryo in a concentration-dependent manner. *Cell* 54: 95–104.

Dubos, R. 1965. *Man Adapting.* Yale University Press, New Haven.

Dunbar, R. I. M. 1992. Neocortex size as a constraint on group size in primates. *Journal of Human Evolution* 20: 469–93.

Durrant, A. 1962. The environmental induction of heritable change in *Linum. Heredity* 17: 27–61.

Dyson, F. J. 1985. *Origins of Life.* Cambridge University Press, Cambridge.

Eichberg, J., Sherwood, E., Epps, D. E. & Oró, J. 1977. Cyanamide mediated synthesis under plausible primitive Earth conditions IV. The synthesis of acyglycerols. *Journal of Molecular Evolution* 10: 211–30.

Eigen, M. 1971. Self-organization of matter and the evolution of biological macro-molecules. *Naturwissenschaften* 58: 465–523.

Eigen, M. & Schuster, P. 1977. The hypercycle. A principle of natural self-organization Part A: emergence of the hypercycle. *Naturwissenschaften* 64: 541–65.

Eigen, M., Schuster, P., Gardiner, W. & Winkler-Oswatitsch, R. 1981. The origin of genetic information. *Scientific American* 244(4): 78–94.

Eigen, M., Lindemann, B. F., Tietze, M., Winkler-Oswatitsch, R., Dress, A. & Haeseler, A. v. 1989. How old is the genetic code? Statistical geometry of tRNA provides an answer. *Science* 244: 673–9.

Ellington, A. D. & Szostak, J. W. 1990. *In vitro* selection of RNA molecules that bind specific ligands. *Nature* 346: 818–22.

Engebrecht, J., Hirsch, J. & Roeder, G. S. 1990. Meiotic gene conversion and crossing over: their relationship to each other and to chromosome synapsis and segregation. *Cell* 62: 927–37.

Engels, W. R. 1989. P elements in *Drosophila*. In: D. Berg & M. Howe (eds), *Mobile DNA*. ASM Publications, Washington, DC.

Epps, D. E., Sherwood, E., Eichberg, J. E. & Oró, J. 1978. Cyanamide mediated synthesis under plausible primitive Earth conditions V. The synthesis of phosphatidic acids. *Journal of Molecular Evolution* 11: 279–92.

Eshel, I. 1985. Evolutionary genetic stability of Mendelian segregation and the role of free recombination in the chromosomal system. *American Naturalist* 125: 412–20.

Ewald, P. W. 1983. Host–parasite relations, vectors, and the evolution of disease severity. *Annual Review of Ecology and Systematics* 14: 465–85.

Famulok, M. 1994. Molecular recognition of amino acids by RNA-aptamers: an L-citrulline binding RNA motif and its evolution into an L-arginine binder. *Journal of the American Chemical Society* (in press).

Ferat, J.-L. & Michel, F. 1993. Group II self-splicing introns in bacteria. *Nature* 364: 358–61.

Fisher, R. A. 1930. *The Genetical Theory of Natural Selection.* Oxford University Press, Oxford.

Fisher, R. A. 1931. The evolution of dominance. *Biological Reviews* 6: 345–86.

Fox, S. W. 1984. Proteinoid experiments and evolutionary theory. In: M. W. Ho & P. T. Saunders (eds), *Beyond Neo-Darwinism*, pp. 15–60. Academic Press, New York.

Fox, S. W. & Dose, K. 1977. *Molecular Evolution and the Origin of Life.* Marcel Dekker, New York.

Frank, S. A. & Crespi, B. J. 1989. Synergism between sib-rearing and sex ratio in Hymenoptera. *Behavioural Ecology and Sociobiology* 24: 155–62.

French, V., Bryant, P. J. & Bryant, S. V. 1976. Pattern regulation of epimorphic fields. *Science* 193: 969–81.

Gánti, T. 1974. Theoretical deduction of the function and structure of the genetic material. *Biológia* 22: 17–35 [in Hungarian].

Gánti, T. 1975. Organisation of chemical reactions into dividing and metabolizing units: the chemotons. *Biosystems* 7: 189–95.

Gánti, 1978. Chemical systems and supersystems III. Models of self-reproducing chemical supersystems: the chemotons. *Acta Chimica Academiae Scientiarum Hungarica* 98: 265–83.

Gánti, T. 1979. *A Theory of Biochemical Supersystems and its Application to Problems of Natural and Artificial Biogenesis.* Akadémiai Kiadó, Budapest; University Park Press, Baltimore.

Gánti, 1983. The origin of the earliest sequences. *Biológia* 31: 47–54 [in Hungarian].

Gánti, T. 1984. *Chemoton Theory I. Theoretical Foundations of Fluid Machineries.* OMIKK, Budapest [in Hungarian].

Gánti, T. 1987. *The Principle of Life.* OMIKK, Budapest.

Garcia-Fernàndez, J. & Holland, P. W. H. 1994. Archetypal organization of the amphioxus *Hox* gene cluster. *Nature* 370: 563–6.

Gardner, B. T. & Gardner, A. R. 1974. Comparing the early utterances of child and chimpanzee. In: A. Pick (ed.), *Minnesota Symposium on Child Psychology*, pp. 3–23. University of Minnesota Press, Minneapolis.

Gause, G. F. 1934. *The Struggle for Existence.* Williams & Wilkins, Baltimore.

Gellner, E. 1988. Origins of society. In: A. C. Fabian (ed.), *Origins*, pp. 128–40. Cambridge University Press, Cambridge.

Gilbert, W. 1978. Why genes in pieces? *Nature* 271: 501.

Go, M. 1991. Module organization in proteins and exon shuffling. In: S. Ozawa & T. Honjo (eds), *Evolution of Life: Fossils, Molecules, and Culture*, pp. 109–22. Springer-Verlag, Tokyo.

Goody, J. 1977. *The Domestication of the Savage Mind.* Cambridge University Press, Cambridge.

Gopnik, M. 1990. Feature-blind grammar and dysphasia. *Nature* 344: 715.

Gopnik, M. & Crago, M. B. 1991. Familial aggregation of a developmental language disorder. *Cognition* 39: 1–50.

Gould, S. J. & Lewontin, R. C. 1979. The spandrels of San Marco and the Panglossian paradigm: a critique of the adaptationist programme. *Proceedings of the Royal Society of London*, B205: 281–8.

Grafen, A. 1985. A geometric view of relatedness. *Oxford Surveys in Evolutionary Biology* 2: 28–89.

Grafen, A. 1986. Split sex ratios and the evolutionary origin of eusociality. *Journal of Theoretical Biology* 122: 95–121.

References

Grassé, P. P. & Noirot, C. 1958. Construction et architecture chez les termites champignonnistes (Macrotermitinae). *Proceedings 10th International Congress of Entomology 1956* 2: 515–20.

Gray, M. W. 1989. The evolutionary origins of organelles. *Trends in Genetics* 5: 294–9.

Gray, M. W. & Doolittle, W. F. 1982. Has the endosymbiotic hypothesis been proven? *Microbiological Reviews* 46: 369–90.

Green, J. B. A. & Smith, J. C. 1990. Graded changes in dose of a *Xenopus* activin A homologue elicit stepwise transitions in embryonic cell fate. *Nature* 347: 391–4.

Greenfield, P. M. 1991. Language, tools and brain: the ontogeny and phylogeny of hierarchically organized sequential behavior. *Behavioral Brain Science* 14: 531–95.

Greenfield, P. M. & Savage-Rumbaugh, E. S. 1991. Imitation, grammatical development, and the invention of protogrammar. In: N. Krasnegor, D. M. Rumbaugh, M. Studdert-Kennedy & R. L. Scheifelbusch (eds), *Biological and Behavioral Determinants of Language Development*, pp. 235–58. Erlbaum: Hillsdale, New Jersey.

Grime, J. P. 1983. Prediction of weed and crop response to climate based upon measurements of nuclear DNA content. In: *Aspects of Biology 4. Influence of Environmental Factors on Herbicide Performance and Crop and Weed Biology*, pp. 87–98.

Grun, P. 1976. *Cytoplasmic Genetics and Evolution*. Columbia University Press, New York.

Guerrier-Takada, C., Gardiner, T., Marsh, N., Pace, S. & Altman, S. 1983. The RNA moiety of ribonuclease P is the catalytic subunit of the enzyme. *Cell* 35: 849–57.

Hadorn, E. 1965. Problems of determination and transdetermination. *Brookhaven Symposia in Biology* 18: 148–61.

Haig, D. 1993. Alternatives to meiosis: the unusual genetics of red algae, microsporidia and others. *Journal of Theoretical Biology* 163: 15–21.

Haig, D. & Grafen, A. 1991. Genetic scrambling as a defence against meiotic drive. *Journal of Theoretical Biology* 153: 531–58.

Haig, D. & Hurst, L. 1991. A quantitative measure of error minimization of the genetic code. *Journal of Molecular Evolution* 33: 412–17.

Haldane, J. B. S. 1929. The origin of life. *Rationalist Annual* 1929: 148–69.

Haldane, J. B. S. 1955. Population genetics. *Penguin New Biology* 15: 44.

Hall, J. L., Ramanis, Z. & Luck, D. J. L. 1989. Basal body/centriolar DNA: molecular genetic studies in *Chlamydomonas*. *Cell* 59: 121–32.

Hamilton, W. D. 1964. The genetical evolution of social behaviour. *Journal of Theoretical Biology* 7: 1–52.

Hamilton, W. D. 1967. Extraordinary sex ratios. *Science* 156: 477–88.

Hamilton, W. D. 1980. Sex versus non-sex versus parasite. *Oikos* 35: 282–90.

Hardin, J. W. 1975. Hybridisation and introgression in *Quercus alba*. *Journal of the Arnold Arboretum, Harvard University* 56: 336–63.

Harold, F. M. 1990. To shape a cell: an inquiry into the causes of morphogenesis of microorganisms. *Microbiological Reviews* 54: 381–431.

Harrison, H. 1985. *West of Eden*. Pantheon Books, London.

Hartl, D. L. & Hiraizumi, Y. 1976. Segregation distortion. In: M. Ashburner & E. Novitski (eds), *Genetics and Biology of Drosophila*, I, pp. 615–66. Academic Press, New York.

Hartmann, H. 1975. Speculations on the origin and evolution of metabolism. *Journal of Molecular Evolution* 4: 359–70.

Hawksworth, D. L. 1988. Coevolution of fungi with algae and cyanobacteria in lichen symbioses. In: Pirozynski & Hawksworth (op. cit.), pp. 125–48.

Heijne, G. 1986. Why mitochondria need a genome. *FEBS Letters* 198: 1–4.

Helfrich, W. 1973. Elastic properties of lipid bilayers: theory and possible experiments. *Zeitschrift für Naturforschung* 28c: 693–703.

Herman, L. M., Richards, D. G. & Wolz, J. P. 1984. Comprehension of sentences by bottlenosed dolphins. *Cognition* 16: 129–219.

Herre, E. A. 1993. Population structure and the evolution of virulence in nematode parasites of fig wasps. *Science* 259: 1442–5.

Herskowitz, I. 1988. Life cycle of the budding yeast *Saccharomyces cerevisiae*. *Microbiological Reviews* 52: 536–53.

Hewitt, G. M. 1973. Evolution and maintenance of B-chromosomes. *Chromosomes Today* 4: 351–69.

Hickey, D. A. & Rose, M. R. 1988. The role of gene transfer in the evolution of sex. In: R. E. Michod & B. R. Levin (eds), *The Evolution of Sex: An Examination of Current Ideas*, pp. 161–93. Sinauer, Sunderland, Mass.

Hinkle, G. 1991. Status of the theory of the symbiotic origin of undulipodia (cilia). In: Margulis & Fester (op. cit.), pp. 135–42.

Hinton, G. E. & Nowlan, S. J. 1987. How learning can guide evolution. *Complex Systems* 1: 495–502.

Hiraga, S. 1993. Chromosome partition in *Escherichia coli*. *Current Opinion in Genetics and Development* 3: 789–801.

Ho, C. K. 1988. Primitive ancestry of transfer RNA. *Nature* 333: 24.

Hoekstra, R. F. 1982. On the asymmetry of sex: evolution of mating types in isogamous populations. *Journal of Theoretical Biology* 87: 785–93.

Hoekstra, R. F. 1987. The evolution of sexes. In S. C. Stearns (ed.), *The Evolution of Sex and its Consequences*, pp. 59–91. Birkhauser, Basel.

Hoekstra, R. F. 1990. The evolution of male–female dimorphism: older than sex? *Journal of Genetics* 69: 11–15.

Hollande, A. & Carruette-Valentin, J. 1970. Appariement chromosomique et complexes synaptomatiques dans les noyaux en cour de depolyploidisation chez *Pyrsonympha flagellata*: le cycle evolutif des Pyrsonymphines symbiontes de *Reticulotermes lucifugus*. *Compte Rendus de l'Acadèmie des Sciences, Paris* D270: 2550–3.

Holliday, R. 1987. The inheritance of epigenetic defects. *Science* 238: 163–70.

Holman, E. W. 1987. Recognizability of sexual and asexual species of rotifers. *Systematic Zoology* 36: 381–6.

Hopf, F. A. & Hopf, F. W. 1985. The role of the Allee effect in species packing. *Theoretical Population Biology* 27: 27–50.

Hurst, L. D. 1990. Parasite diversity and the evolution of diploidy, multicellularity and anisogamy. *Journal of Theoretical Biology* 144: 429–43.

Hurst, L. D. 1993. The incidences, mechanisms and evolution of cytoplasmic sex ratio distorters in animals. *Biological Reviews* 68: 121–94.

Hurst, L. D. & Hamilton, W. D. 1992. Cytoplasmic fusion and the nature of the sexes. *Proceedings of the Royal Society of London* B247: 189–94.

Hurst, L. D. & Nurse, P. 1991. A note on the evolution of meiosis. *Journal of Theoretical Biology* 150: 561–3.

Hurst, L. D. & Pomiankowski, A. N. 1991a. Causes of sex ratio bias within species may also account for unisexual sterility in species hybrids: a new explanation of Haldane's rule and related phenomena. *Genetics* 128: 841–58.

Hurst, L. D. & Pomiankowski, A. N. 1991b. Maintaining Mendelism: prevention is better than cure. *BioEssays* 13: 489–90.

Ingham, P. W. 1988. The molecular genetics of embryonic pattern formation in *Drosophila*. *Nature* 335: 24–34.

Jablonka, E. & Lamb, M. J. 1989. The inheritance of acquired epigenetic variations. *Journal of Theoretical Biology* 139: 69–83.

Jablonka, E., Lachmann, M. & Lamb, M. J. 1992. Evidence, mechanisms and models for the inheritance of acquired characters. *Journal of Theoretical Biology* 158: 245–68.

Jackendoff, R. 1977. *X-bar Syntax: A Study of Phrase Structure*. MIT Press, Cambridge, Mass.

Jacob, F. & Monod, J. 1961, On the regulation of gene activity. *Cold Spring Harbor Symposia on Quantitative Biology* 26: 193–211.

Jacob, F., Brenner, S. & Cutzin, F. 1963. On the regulation of DNA replication in bacteria. *Cold Spring Harbor Symposia on Quantitative Biology* 28: 329–48.

Johnson, K. A. & Rosenbaum, J. L. 1991. Basal bodies and DNA. *Trends in Cell Biology* 1: 145–9.

Jones, R. N. & Rees, H. 1982. *B-Chromosomes*. Academic Press, New York.

Jones, R. N. 1975. B chromosome systems in flowering plants and animal species. *International Review of Cytology* 40: 1–100.

Joyce, G. F., Schwartz, A. W., Orgel, L. E. & Miller, S. L. 1987. The case for an ancestral genetic system involving simple analogues of the nucleotides. *Proceedings of the National Academy of Sciences, USA* 84: 4398–402.

Joyce, G. G. 1989. RNA evolution and the origins of life. *Nature* 338: 217–24.

Jukes, T. H. & Osawa, S. 1991. Recent evidence for the evolution of the genetic code. In: S. Osawa & T. Honjo (eds), *Evolution of Life: Fossils, Molecules, and Culture*, pp. 79–95. Springer-Verlag, Tokyo.

Jukes, T. H., Holmquist, R. & Moise, H. 1975. Amino acid composition of proteins: selection against the genetic code. *Science* 189: 50–1.

Kacser, H. & Beeby, R. 1984. Evolution of catalytic proteins: or, on the origin of enzyme species by means of natural selection. *Journal of Molecular Evolution* 20: 38–51.

Kacser, H. & Burns, J. A. 1973. The control of flux. *Symposia of the Society for Experimental Biology* 32: 65–104.

Kacser, H. & Burns, J. A. 1979. Molecular democracy: who shares the controls? *Biochemical Society Transactions* 7: 1149–60.

Kacser, H. & Burns, J. A. 1981. The molecular basis of dominance. *Genetics* 97: 639–66.

Kaiser, D. 1986. Control of multicellular development: *Dictyostelium* and *Myxococcus*. *Annual Review of Genetics* 20: 539–66.

Kaler, E. W., Murthy, A. K., Rodriguez, B. E. & Zasadzinski, J. A. N. 1989. Spontaneous vesicle formation in aqueous mixtures of single-tailed surfactants. *Science* 245: 1371–4.

Kasting, J. F. & Ackerman, T. P. 1986. Climatic consequences of very high carbon dioxide levels in the Earth's early atmosphere. *Science* 234: 1383–5.

Kauffman, S. A. 1986. Autocatalytic sets of proteins. *Journal of Theoretical Biology* 119: 1–24.

Kaufman, M. H. 1972. Non-random segregation during mammalian oogenesis. *Nature* 238: 465–6.

Kazakov, S. & Altman, S. 1992. A trinucleotide can promote metal ion-dependent specific cleavage of RNA. *Proceedings of the National Academy of Sciences, USA* 89: 7939–43.

Kendrick, B. 1991. Fungal symbioses and evolutionary innovation. In: Margulis & Fester (op. cit.), pp. 249–61.

Kiedrowski, G. von 1986. A self-replicating hexadeoxy nucleotide. *Angewandte Chemie, International Edition (English)* 25: 932–5.

Kiedrowski, G. von 1993. Minimal replicator theory I: parabolic versus exponential growth. *Bioorganic Chemistry Frontiers* 3: 113–46.

King, G. A. M. 1980. Evolution of the coenzymes. *Biosystems* 13: 23–45.

King, G. A. M. 1982. Recycling, reproduction and life's origins. *Biosystems* 15: 89–97.

King, G. A. M. 1986. Was there a prebiotic soup? *Journal of Theoretical Biology* 123: 493–8.

Kirk, D. L. 1988. The ontogeny and phylogeny of cellular differentiation in *Volvox*. *Trends in Genetics* 4: 32–6.

Kirk, D. L. & Harper, J. F. 1986. Genetic, biochemical, and molecular approaches to *Volvox* development and evolution. *International Review of Cytology* 99: 217–93.

Knowlton, N. 1974. A note on the evolution of gamete dimorphism. *Journal of Theoretical Biology* 46: 283–5.

Koch, A. L. 1985. Primeval cells: possible energy-generating and cell-division mechanisms. *Journal of Molecular Evolution* 21: 270–7.

Kochanski, R. S. & Borisy, G. G. 1990. Mode of centriole duplication and distribution. *Journal of Cell Biology* 110: 1599–605.

Koehler, W. 1925. *The Mentality of Apes*. Harcourt, Brace & World, New York.

Kondrashov, A. S. 1982. Selection against harmful mutations in large sexual and asexual populations. *Genetical Research, Cambridge* 40: 325–32.

Kondrashov, A. S. 1988. Deleterious mutations and the evolution of sexual reproduction. *Nature* 336: 435–40.

Konecny, J., Eckert, E., Schöniger, M. & Hofacker, G. L. 1993. Neutral adaptation of the genetic code to double-strand coding. *Journal of Molecular Evolution* 36: 407–16.

Kozo-Polyanski, B. M. 1924. *New Principles of Biology*. Puchina, Moscow [in Russian].

Kruger, K., Grabowski, P. J., Zaug, A. J., Sands, J., Gottschling, D. E. & Cech, T. R. 1982. Self-splicing RNA: autoexcision and autocyclization of the ribosomal intervening sequences of *Tetrahymena*. *Cell* 31: 147–57.

Kunz, B. A. & Haynes, R. H. 1981. Phenomenology and genetic control of mitotic recombination in yeast. *Annual Review of Genetics* 15: 57–89.

Lake, J. A. & Rivera, M. C. 1994. Was the nucleus the first endosymbiont? *Proceedings of the National Academy of Sciences, USA* 91: 2880–1.

Landmann, O. E. 1991. The inheritance of acquired characteristics. *Annual Review of Genetics* 25: 1–20.

Langley, C. H., Montgomery, E., Hudson, R., Kaplan, N. & Charlesworth, B. 1988. On the role of unequal exchange in the containment of transposable element copy number. *Genetical Research, Cambridge* 52: 223–35.

Law, R. 1991. The symbiotic phenotype: origins and evolution. In: Margulis & Fester (op. cit.), pp. 57–71.

Law, R. & Lewis, D. H. 1983. Biotic environments and the maintenance of sex: some evidence from mutualistic symbioses. *Biological Journal of the Linnean Society* 20: 249–76.

Lawrence, P. A. 1992. *The Making of a Fly*. Blackwell, Oxford.

Lazcano, A., Guerrero, R., Margulis, L. & Oró, J. 1988. The evolutionary transition from RNA to DNA in early cells. *Journal of Molecular Evolution* 27: 283–90.

Leigh, E. G. 1971. *Adaptation and Diversity*. Freeman, Cooper and Co., San Francisco.

Lengyel, I. & Epstein, I. R. 1991. Modeling of Turing structures in the chlorite–iodide–malonic acid–starch reaction system. *Science* 251: 650–2.

Lewis, D. 1941. Male sterility in natural populations of hermaphrodite plants. *New Phytologist* 40: 158–60.

Lewis, D. H. 1991. Mutualistic symbioses in the origin and evolution of land plants. In: Margulis & Fester (op. cit.), pp. 288–300.

Lewis, W. M. Jr. 1985. Nutrient scarcity as an evolutionary cause of haploidy. *American Naturalist* 125: 692–701.

Lieberman, P. 1984. *The Biology and Evolution of Language*. Harvard University Press, Cambridge, Mass.

Lieberman, P. 1989. The origins of some aspects of human language and cognition. In: P. Mellars & C. B. Stringer (eds), *The Human Revolution: Behavioural and Biological Perspectives in the Origins of Modern Humans*, pp. 391–414. Edinburgh University Press, Edinburgh.

Lipowsky, R. 1991. The conformation of membranes. *Nature* 349: 475–81.

Macgregor, H. C. 1982. Big chromosomes and speciation among amphibia. In: G. A. Dover & R. B. Flavell (eds), *Genome Evolution*, pp. 325–41. Academic Press, London.

Magee, T. R., Asai, T., Malka, D. & Kogoma, T. 1992. DNA damage-inducible origins of DNA replication in *Escherichia coli*. *EMBO Journal* 11: 4219–25.

Maher, K. A. & Stevenson, D. J. 1988. Impact frustration of the origin of life. *Nature* 331: 612–14.

Maiagolowkin, C. & Poulson, D. F. 1957. Infective transfer of maternally inherited sex ratio in *Drosophila paulistorum*. *Science* 126: 32.

Maniotis, A. & Schliwa, M. 1991. Microsurgical removal of centrosomes blocks cell reproduction and centriole generation in Bsc-1 cells. *Cell* 67: 495–504.

Margulis, L. 1970. *Origin of Eukaryotic cells*. Yale University Press, New Haven.

Margulis, L. 1981. *Symbiosis in Cell Evolution*. W. H. Freeman, San Francisco.

Margulis, L. 1991. Symbiosis in evolution: origin of cell motility. In: S. Osawa & T. Honjo (eds), *Evolution of Life: Fossils, Molecules, and Culture*, pp. 305–24. Springer-Verlag, Tokyo.

Margulis, L. & Fester, R. (eds) 1991. *Symbiosis as a Source of Evolutionary Innovation*. MIT Press, Cambridge, Mass.

Margulis, L. & Sagan, D. 1986. *Origins of Sex: Three Billion Years of Recombination*. Yale University Press, New Haven.

Margulis, L., Chase, D. & To, L. 1978. Microtubules in prokaryotes. *Science* 200: 1118–23.

Margulis, L., Chase, D. & To, L. P. 1979. Possible evolutionary significance of spirochetes. *Transactions of the Royal Society of London* B204: 189–98.

Martin, W., Somerville, C. C. & Loiseaux-de Goer, S. 1992. Molecular phylogenies of plastid origins and algal evolution. *Journal of Molecular Evolution* 35: 385–404.

Matsuzawa, T. 1991. Nesting cups and metatools in chimpanzees. *Behavioral Brain Science* 14: 570–1.

May, R. M. & Anderson, R. M. 1982. Coevolution of hosts and parasites. *Parasitology* 85: 411–26.

May, R. M. & Anderson, R. M. 1983. Epidemiology and genetics in the coevolution of parasites and hosts. *Proceedings of the Royal Society of London* B219: 281–313.

May, R. M. & Nowak, M. 1992. Evolutionary games and spatial chaos. *Nature* 359: 826–9.

Maynard Smith, J. 1966. Sympatric speciation. *American Naturalist* 100: 637–50.

Maynard Smith, J, 1971. The origin and maintenance of sex. In: G. C. Williams (ed.), *Group Selection*, pp. 163–75. Aldine-Atherton, Chicago.

Maynard Smith, J. 1978. *The Evolution of Sex*. Cambridge University Press, Cambridge.

Maynard Smith, J. 1979a. Hypercycles and the origin of life. *Nature* 20: 445–6.

Maynard Smith, J. 1979b. The effect of normalizing and disruptive selection on genes for recombination. *Genetical Research, Cambridge* 33: 121–8.

Maynard Smith, J. 1983a. Models of Evolution. *Proceedings of the Royal Society of London* B219: 315–25.

Maynard Smith, J. 1983b. Game theory and the evolution of cooperation. In: D. S. Bendall (ed.), *Evolution from Molecules to Man*, pp. 445–56. Cambridge University Press, Cambridge.

Maynard Smith, J. 1988a. Evolutionary progress and the levels of selection. In: M. H.

Nitecki (ed.), *Evolutionary Progress*, pp. 219–30. University of Chicago Press, Chicago.

Maynard Smith, J. 1988b. Selection for recombination in a polygenic model—the mechanism. *Genetical Research, Cambridge* 51: 59–63.

Maynard Smith, J. 1990. Models of a dual inheritance system. *Journal of Theoretical Biology* 143: 41–53.

Maynard Smith, J. 1991. A Darwinian view of symbiosis. In: Margulis & Fester (op. cit.), pp. 26–39.

Maynard Smith, J. & Sondhi, K. C. 1961. The arrangement of bristles in *Drosophila*. *Journal of Embryology and Experimental Morphology* 9: 611–72.

Maynard Smith, J. & Szathmáry, E. 1993. The origin of chromosomes I. Selection for linkage. *Journal of Theoretical Biology* 164: 437–66.

Maynard Smith, J., Dowson, C. G. & Spratt, B. G. 1991. Localized sex in bacteria. *Nature* 349: 29–31.

Mayr, E. 1942. *Systematics and the Origin of Species*. Columbia University Press, New York.

McFall-Ngai, M. J. 1991. Luminous bacterial symbioses in fish evolution: adaptive radiation among the leiognathid fishes. In: Margulis & Fester (op. cit.), pp. 380–409.

Meinhardt, H. 1982. *Models of Biological Pattern Formation*. Academic Press, London.

Mereschowsky, C. 1910. Theorie der Zwei Pflanzenarten als Grundlage der Symbiogenesis, einer neuen Lehre der Entstehung der Organismen. *Biologisches Zentralblatt* 30: 278–303, 321–47, 353–67.

Michod, R. 1983. Population biology of the first replicators: the origin of genotype, phenotype and organism. *American Zoologist* 23: 5–14.

Miller, S. L. 1953. A production of amino acids under possible primitive Earth conditions. *Science* 117: 528–9.

Miller, S. L. 1987. Which organic compounds could have occurred on the prebiotic earth? *Cold Spring Harbor Symposia on Quantitative Biology* 52: 17–27.

Miyata, T., Iwabe, N., Kuma, K.-I., Kawanishi, Y.-I., Hasegawa, M., Kishino, H. *et al.* 1991. Evolution of archaebacteria: phylogenetic relationships among archaebacteria, eubacteria, and eukaryotes. In: S. Osawa & T. Honjo (eds), *Evolution of Life: Fossils, Molecules, and Culture*, pp. 337–51. Springer-Verlag, Tokyo.

Mohr, G., Caprara, M. G., Guo, Q. & Lambowitz, A. M. 1994. A tyrosyl-tRNA synthetase can function similarly to an RNA structure in the *Tetrahymena* ribozyme. *Nature* 370: 147–50.

Molnár, I. 1993. A conception of development and its evolutionary significance. *Abstracta Botanica (Budapest)* 17: 207–24.

Molnár, I. 1994. The division of labour in evolution. *Természet Világa* (in press) [in Hungarian].

Montgomery, E., Charlesworth, B. & Langley, C. H. 1987. A test for the role of natural selection in the stabilisation of transposable element copy number in *Drosophila melanogaster*. *Genetical Research, Cambridge* 49: 31–41.

Muller, C. H. 1951. The oaks of Texas. *Contribution of the Texas Research Foundation* 1: 21–323.

Muller, H. J. 1932. Some genetic aspects of sex. *American Naturalist* 66: 118–38.

Muller, H. J. 1964. The relation of recombination to mutational advance. *Mutation Research* 1: 2–9.

Muller, H. J. 1966. The gene material as the initiator and organizing basis of life. *American Naturalist* 100: 493–517.

Munson, M. A., Baumann, P., Clark, M. A., Baumann, L., Moran, N. A., Voegtlin, D. J. & Campbell, B. C. 1991. Evidence for the establishment of aphid–eubacterium endosymbiosis in an ancestor of four aphid families. *Journal of Bacteriology* 173: 6321–4.

Murray, A. W. 1992. Creative blocks: cell-cycle checkpoints and feedback controls. *Nature* 359: 599–604.

Murray, J. D. 1990. Discussion: Turing's theory of morphogenesis—its influence on modelling biological pattern and form. *Bulletin of Mathematical Biology* 52: 119–52.

Myles, D. G. 1975. Structural changes in the sperm of *Marsilea vestita* before and after fertilization. In: J. G. Duckett & P. A. Racey (eds), *The Biology of the Male Gamete*, *Biological Journal of the Linnean Society* 7, Suppl. 1, pp. 129–34. Academic Press, London.

Nardon, P. & Grenier, A.-M. 1991. Serial endosymbiosis theory and weevil evolution: the role of symbiosis. In: Margulis & Fester (op. cit.), pp. 153–69.

Nealson, K. H. 1991. Luminescent bacteria symbiotic with entomopathogenic nematodes. In: Margulis & Fester (op.cit.), pp. 205–18.

Nee, S. & Maynard Smith, J. 1990. The evolutionary biology of molecular parasites. *Parasitology* 100: S5–S18.

Niesert, V., Harnasch, D. & Bresch, C. 1981. Origin of life between Scylla and Charybdis. *Journal of Molecular Evolution* 17: 348–53.

Noller, H. F., Hoffarth, V. & Zimnniak, L. 1992. Unusual resistance of peptidyl transferase to protein extraction procedures. *Science* 256: 1416–24.

Nowak, M. & Sigmund, K. 1993. A strategy of win-stay, lose-shift that outperforms tit-for-tat in the Prisoner's Dilemma game. *Nature* 364: 56–57.

Nurse, P. 1985. The genetic control of cell volume. In: T. Cavalier-Smith (ed.), *The Evolution of Genome Size*, pp. 185–96. Wiley, Chichester.

Oberbeck, V. R., Marshall, J. & Shen, T. 1991. Prebiotic chemistry in clouds. *Journal of Molecular Evolution* 32: 296–303.

Ochman, H., Whittam, T. S., Caugant, D. A. & Selander, R. K. 1983. Enzyme polymorphism and genetic population structure in *Escherichia coli* and *Shigella*. *Journal of General Microbiology* 129: 2715–26.

Ohama T., Osawa, S., Watanabe, K. & Jukes, T. H. 1990. Evolution of the mitochondrial genetic code IV. AAA is an asparagine codon in some animal mitochondria. *Journal of Molecular Evolution* 30: 329–32.

Ohno, S. 1987. Evolution from primordial oligomeric repeats to modern coding sequences. *Journal of Molecular Evolution* 25: 325–9.

Ono, B. *et al.* 1990. Alternative self-diploidization or 'ASD' homothallism in *Saccharomyces*

cerevisiae: isolation of a mutant, nuclear-cytoplasmic interaction and endomitotic diploidization. *Genetics* 125: 729–38.

Oparin, A. I. 1924. *Proiskhozhdenie Zhizny* [*The Origin of Life*]. Izd. 'Moskovskiy Rabochiy', Moscow.

Orgel, L. E. 1963. The maintenance of the accuracy of protein synthesis and its relevance to ageing. *Proceedings of the National Academy of Sciences, USA* 49: 517–21.

Orgel, L. E. 1968. Evolution of the genetic apparatus. *Journal of Molecular Biology* 38: 381–93.

Orgel, L. E. 1979. Selection *in vitro*. *Proceedings of the Royal Society of London* B205: 435–42.

Orgel, L. E. 1987. Evolution of the genetic apparatus: a review. *Cold Spring Harbor Symposia on Quantitative Biology* 52: 9–16.

Orgel, L. E. 1989. The origin of polynucleotide-directed protein synthesis. *Journal of Molecular Evolution* 29: 465–74.

Orgel, L. E. 1990. Adding to the genetic alphabet. *Nature* 343: 18–20.

Orgel, L. E. & Crick, F. H. C. 1980. Selfish DNA: the ultimate parasite. *Nature* 284: 604–7.

Oró, J. 1961. Comets and the formation of biochemical compounds on the primitive Earth. *Nature* 190: 389.

Osawa, S. & Jukes, T. H. 1988. Evolution of the genetic code as affected by anticodon content. *Trends in Genetics* 4: 191–8.

Osawa, S. & Jukes, T. H. 1989. Codon reassignment (codon capture) in evolution. *Journal of Molecular Evolution* 28: 271–8.

Osawa, S., Jukes, T. H., Wanatabe, K. & Muti, A. 1992. Recent evidence for evolution of the genetic code. *Microbiological Reviews* 56: 229–64.

Oster, G. F. & Wilson, E. O. 1978. *Caste and Ecology in the Social Insects*. Princeton University Press, Princeton.

Owen, D. F. 1970. Inheritance of the sex ratio in the butterfly *Acraea encedon*. *Nature* 225: 662–3.

Packer, C. & Ruttan, L. 1988. The evolution of cooperative hunting. *American Naturalist* 132: 159–98.

Packer, C., Herbst, L., Pusey, A. E., Bygott, J. D., Cairns, S. J. & Borgerhoff-Mulder, M. 1985. Reproductive success in lions. In: T. H. Clutton-Brock (ed.), *Reproductive Success*, pp. 363–83. University of Chicago Press, Chicago.

Packer, L. 1986. Multiple-foundress associations in a temperate population of *Halictis ligatus* (Hymenoptera, Halictidae). *Canadian Journal of Zoology* 64: 2325–32.

Page, J. R. Jr & Robinson, G. E. 1991. The genetics of division of labour in honeybees. *Advances in Insect Physiology* 23: 117–69.

Palmer, J. D. 1993. A genetic rainbow of plastids. *Nature* 364: 762–3.

Palmer, J. D. & Logsdon, J. M. 1991. The recent origin of introns. *Current Opinion in Genetics and Development* 1: 470–7.

Parker, G. A. & MacNair, M. R. 1978. Models of parent–offspring conflict, I. Monogamy. *Animal Behaviour* 26: 97–110.

Parker, G. A., Baker, R. R. & Smith, V. G. F. 1972. The origin and evolution of gamete dimorphism and the male–female phenomenon. *Journal of Theoretical Biology* 36: 529–53.

Penrose, L. S. 1959. Self-reproducing machines. *Scientific American* 200(6): 105–14.

Peterson, S. P. & Berns, M. W. 1980. The centriolar complex. *International Review of Cytology* 64: 82–106.

Piaget, J. 1952. *The Origins of Intelligence in Children*. International University Press.

Piatelli-Palmarini, M. 1989. Evolution, selection and cognition: from 'learning' to parameter setting in biology and the study of language. *Cognition* 31: 1–44.

Piccirilli, J. A., Krauch, T., Moroney, S. E. & Benner, S. A. (1990). Enzymatic incorporation of a new base pair into DNA and RNA extends the genetic alphabet. *Nature* 343: 33–37.

Piccirilli, J. A., McConnell, T. S., Zaug, A. J., Noller, H. F. & Cech, T. R. 1992. Aminoacyl esterase activity of the *Tetrahymena* ribozyme. *Science* 256: 1420–4.

Pimm, S. L. & Lawton, J. H. 1977. The number of trophic levels in ecological communities. *Nature* 268: 329–31.

Pinker, S. & Bloom, P. 1990. Natural language and natural selection. *Behavioral Brain Science* 13: 707–84.

Pirozynski, K. A. & Hawksworth, D. L. (eds) 1988. *Coevolution of Fungi with Plants and Animals*. Academic Press, London.

Plooij, F. X. 1978. Some basic traits of language in wild chimpanzees. In: A. W. Lock (ed.), *Action, Gesture, and Symbol: The Emergence of Language*, pp. 111–31. Academic Press, New York.

Pollack, S. J., Jacobs, J. W. & Schultz, P. G. 1986. Selective chemical catalysis by an antibody. *Science* 234: 1570–3.

Pool, R. 1990. The third kingdom of life. *Science* 247: 158–60.

Premack, D. 1985. 'Gavagai!' or the future history of the animal language controversy. *Cognition* 19: 207–96.

Prudent, J. R., Uno, T. & Schultz, P. G. 1994. Expanding the scope of RNA catalysis. *Science* 264: 1924–7.

Raikov, I. B. 1982. *The Protozoan Nucleus*. Springer-Verlag, Heidelberg and New York.

Ramanis, Z. & Luck, D. J. L. 1986. Loci affecting flagellar assembly and function map to an unusual linkage group in *Chlamydomonas reinhardtii*. *Proceedings of the National Academy of Sciences, USA* 83: 423–6.

Rashevsky, N. 1938. *Mathematical Biophysics*. University of Chicago Press, Chicago.

Ratnieks, F. L. W. 1988. Reproductive harmony via mutual policing by workers in eusocial Hymenoptera. *American Naturalist* 132: 217–36.

Ratnieks, F. L. W. & Visscher, P. K. 1989. Worker policing in the honeybee. *Nature* 342: 796–7.

Raven, P. 1970. A multiple origin for plastids and mitochondria. *Science* 169: 641–6.

Rebek, J. 1994. Synthetic self-replicating molecules. *Scientific American* 271(1): 34–40.

Robinson, G. E. 1987. Regulation of honey bee age polyethism by juvenile hormone. *Behavioural Ecology and Sociobiology* 20: 329–38.

Rodin, S., Ohno, S. & Rodin, A. 1993. Transfer RNAs with complementary anticodons: could they reflect early evolution of discriminative genetic code adaptors? *Proceedings of the National Academy of Sciences, USA* 90: 4723–7.

Rogers, J. H. 1990. The role of introns in evolution. *FEBS Letters* 268: 339–43.

Roth, G., Rottluff, B. & Linke, R. 1988. Miniaturization, genome size and the origin of functional constraints in the visual system of salamanders. *Naturwissenschaften* 75: 297–304.

Roughgarden, J. A. 1979. *Theory of Population Genetics and Evolutionary Ecology: An Introduction*. Macmillan, New York.

Safran, S. A., Pincus, P. & Andelman, D. 1990. Theory of spontaneous vesicle formation in surfactant mixtures. *Science* 248: 354–6.

Sager, R. 1977. Genetic analysis of chloroplast DNA in *Chlamydomonas*. *Advances in Genetics* 19: 287–340.

Sahlins, M. 1976. *The Use and Abuse of Biology*. University of Michigan Press, Ann Arbor.

Schwartz, A. W. & Orgel, L. E. 1985. Template-directed synthesis of novel, nucleic acid-like structures. *Science* 228: 585–7.

Seger, J. 1983. Partial bivoltinism may cause alternating sex-ratio biases that favour eusociality. *Nature* 301: 59–62.

Seger, J. 1991. Cooperation and conflict in social insects. In: J. R. Krebs & N. B. Davies (eds), *Behavioural Ecology*, 3rd edn, pp. 338–73. Blackwell, Oxford.

Seilacher, A. 1992. Vendobionta and psammocorallia—lost constructions of Precambrian evolution. *Journal of the Geological Society of London* 149: 607–13.

Seyfarth, R. M. & Cheney, D. L. 1992. Meaning and mind in monkeys. *Scientific American* 267(6): 122–8.

Sharp, P. M. 1991. Determinants of DNA sequence divergence between *Escherichia coli* and *Salmonella typhimurium*—codon usage, map position and concerted evolution. *Journal of Molecular Evolution* 33: 23–33.

Shih, M.-C., Heinrich, P. & Goodman, H. M. 1988. Intron existence predated the divergence of eukaryotes and prokaryotes. *Science* 242: 1164–6.

Shimizu, M. 1982. Molecular basis for the genetic code. *Journal of Molecular Evolution* 18: 297–303.

Shimkets, L. J. 1990. Social and developmental biology of myxobacteria. *Microbiological Reviews* 54: 473–501.

Shvedova, T. A., Korneeva, G. A., Otroshchenko, V. A. & Venkstern, T. V. 1987. Catalytic activity of the nucleic acid component of the 1,4-α-glucan branching enzyme from rabbit muscles. *Nucleic Acid Research* 14: 1745–52.

Silver, L. M. 1985. Mouse t haplotypes. *Annual Review of Genetics* 19: 179–208.

Slack, J. M. W., Holland, P. W. H. & Graham, C. F. 1993. The zootype and the phylotypic stage. *Nature* 361: 490–2.

Smith, D. A. S. 1975. All-female broods in the polymorphic butterfly *Danaus chrysippus* L. and their ecological significance. *Heredity* 34: 363–71.

Smith, M. W., Feng, D.-F. & Doolittle, R. F. 1992. Evolution by acquisition: the case for horizontal gene transfers. *Trends in Biochemical Sciences* 17: 489–93.

Smithies, D. & Powers, P. A. 1986. Gene conversions and their relationship to homologous pairing. *Philosophical Transactions of the Royal Society of London* B312: 291–302.

Sogin, M. L. 1991. Early evolution of eukaryotes. *Current Opinion in Genetics and Development* 1: 457–63.

Sogin, M. L., Gundersdon, J. H., Elwood, H. J., Alonso, R. A. & Peattie, D. A. 1989. Phylogenetic meaning of the kingdom concept: an unusual ribosomal RNA from *Giardia lamblia. Science* 243: 75–7.

Sonea, S. 1991. Bacterial evolution without speciation. In: Margulis & Fester (op. cit.), pp. 95–105.

Sonea, S. & Panisset, M. 1983. *A New Bacteriology.* Jones & Bartlett, Boston, Mass.

Sonneborn, T. M. 1965. Degeneracy of the genetic code: extent, nature, and genetic implications. In: V. Bryson & J. H. Vogel (eds), *Evolving Genes and Proteins*, pp. 377–9. Academic Press, New York.

Sonnenfeld, E. M., Koch, A. L. & Doyle, R. J. 1985. Cellular location of origin and terminus of replication in *Bacillus subtilis. Journal of Bacteriology* 163: 895–9.

Spiegleman, S. 1970. Extracellular evolution of replicating molecules. In: F. O. Schmitt (ed.), *The Neuro Sciences: A Second Study Program*, pp. 927–45. Rockefeller University Press, New York.

Stern, D. B. & Lonsdale, D. M. 1982. Mitochondrial and chloroplast genomes of maize have a 12-kilobase DNA sequence in common. *Nature* 299: 698–702.

Stilwell, W. 1976. Facilitated diffusion of amino acids across biomolecular lipid membranes as a model for selective accumulation of amino acids in a primordial protocell. *Biosystems* 8: 111–17.

Stoltzfus, A., Spencer, D. F., Zuker, M., Logsdon, J. M. & Doolittle, W. F. 1994. Testing the exon theory of genes: the evidence from protein structure. *Science* 265: 202–7.

Stouthamer, R., Luck, R. F. & Hamilton, W. D. 1990. Antibiotics cause parthenogenetic *Trichogramma* (Hymenoptera, Trichogrammatidae) to revert to sex. *Proceedings of the National Academy of Sciences, USA* 87: 2424–7.

Stringer, C. & Gamble, C. 1993. *In Search of the Neanderthals: Solving the Puzzle of Human Origins.* Thames and Hudson, London.

Stubblefield, J. W. & Charnov, E. L. 1986. Some conceptual issues in the origin of eusociality. *Heredity* 57: 181–7.

Sugiyama, Y. & Koman, J. 1979. Tool-using and -making behavior in wild chimpanzees in Bossou, Guinea. *Primates* 20: 513–24.

Sumper, M. & Luce, R. 1975. Evidence for *de novo* production of self-replicating and environmentally adapted RNA structures by bacteriophage Qβ replicase. *Proceedings of the National Academy of Sciences, USA*: 72: 162–6.

Swanson, R. 1984. A unifying concept for the amino acid code. *Bulletin of Mathematical Biology* 46: 187–203.

Swetina, J. & Schuster, P. 1982. A model for polynucleotide replication. *Biophysical Chemistry* 16: 329–45.

Szathmáry, E. 1984. The roots of individual organization. In: G. Vida (ed.), *Evolution IV: Frontiers of Evolution Research*, pp. 37–157. Natura, Budapest [in Hungarian].

Szathmáry, E. 1986. The eukaryotic cell as an information integrator. *Endocytobiological Cell Research* 3: 113–32.

Szathmáry, E. 1987. Early evolution of microtubules and undulipodia. *Biosystems* 20: 115–31.

Szathmáry, E. 1989a. The integration of the earliest genetic information. *Trends in Ecology and Evolution* 4: 200–4.

Szathmáry, E. 1989b. The emergence, maintenance, and transitions of the earliest evolutionary units. *Oxford Surveys in Evolutionary Biology* 6: 169–205.

Szathmáry, E. 1990a. Towards the evolution of ribozymes. *Nature* 344: 115.

Szathmáry, E. 1990b. Coding coenzyme handles: useful coding before translation. In: B. Lukács *et al.* (eds), *Evolution: From Cosmogenesis to Biogenesis*, pp. 77–83. KFKI-1990-50/C reprint. Budapest.

Szathmáry, E. 1991a. Simple growth laws and selection consequences. *Trends in Ecology and Evolution* 6: 366–70.

Szathmáry, E. 1991b. Four letters in the genetic alphabet: a frozen evolutionary optimum? *Proceedings of the Royal Society of London* B245: 91–9.

Szathmáry, E. 1991c. Codon swapping as a possible evolutionary mechanism. *Journal of Molecular Evolution* 32: 178–82.

Szathmáry, E. 1992. What is the optimum size for the genetic alphabet? *Proceedings of the National Academy of Sciences, USA* 89: 2614–18.

Szathmáry, E. 1993a. Coding coenzyme handles: a hypothesis for the origin of the genetic code. *Proceedings of the National Acadamy of Sciences, USA* 90: 9916–20.

Szathmáry, E. 1993b. Do deleterious mutations act synergistically? Metabolic control theory provides a partial answer. *Genetics* 133: 127–32.

Szathmáry, E. 1994. Toy models for simple forms of multicellularity, soma and germ. *Journal of Theoretical Biology* 169: 125–32.

Szathmáry, E. & Demeter, L. 1987. Group selection of early replicators and the origin of life. *Journal of Theoretical Biology* 128: 463–86.

Szathmáry, E. & Gladkih, I. 1989. Coexistence of non-enzymatically replicating templates. *Journal of Theoretical Biology* 138: 55–8.

Szathmáry, E. & Maynard Smith, J. 1993a. The origin of chromosomes II. Molecular mechanisms. *Journal of Theoretical Biology* 164: 447–54.

Szathmáry, E. & Maynard Smith, J. 1993b. The origin of genetic systems. *Abstracta Botanica (Budapest)* 17: 197–206.

Szathmáry, E., Scheuring, I., Kotsis, M. & Gladkih, I. 1990. Sexuality of eukaryotic unicells: hyperbolic growth, coexistence of facultative parthenogens, and the repair hypothesis. In: J. Maynard Smith & G. Vida (eds), *Organizational Constraints on the Dynamics of Evolution*, pp. 279–87. Manchester University Press, Manchester.

Szathmáry, E. & Zinztaras, E. 1992. A statistical test of hypotheses on the organisation and origin of the genetic code. *Journal of Molecular Evolution* 35: 185–9.

Tarumi, K. & Schwegler, H. 1987. A non-linear treatment of the protocell model by a boundary layer approximation. *Bulletin of Mathematical Biology* 47: 307–20.

Taylor, F. J. R. & Coates, D. 1989. The code within the codons. *Biosystems* 22: 177–87.

Thomas, C. A. 1971. The genetic organisation of chromosomes. *Annual Review of Genetics* 5: 237–56.

Thomson, K. S. 1972. An attempt to reconstruct evolutionary changes in the cellular DNA content of lungfish. *Journal of Experimental Zoology* 180: 363–72.

Tooby, J. & Cosmides, L. 1990. Toward an adaptationist psycholinguistics. *Behavioral Brain Science* 13: 760–2.

Tramontano, A., Janda, K. D. & Lerner, R. A. 1986. Catalytic antibodies. *Science* 234: 1566–70.

Trench, R. K. 1991. *Cyanophora paradoxa* and the origins of chloroplasts. In: Margulis & Fester (op. cit.), pp. 143–50.

Trivers, R. L. 1974. Parent–offspring conflict. *American Zoologist* 14: 249–64.

Trivers, R. L. & Hare, H. 1976. Haplodiploidy and the evolution of the social insects. *Science* 191: 249–63.

Tuerk, C. & Gold, L. 1990. Systematic evolution of ligands by exponential enrichment: RNA ligands to bacteriophage T4 DNA polymerase. *Science* 249: 505–10.

Turing, A. M. 1952. The chemical basis of morphogenesis. *Philosophical Transactions of the Royal Society of London* B237: 37–72. Reprinted in *Bulletin of Mathematical Biology* 52: 153–97 (1990).

Van der Woude, M. W., Braaten, B. A. & Low, D. A. 1992. Evidence of global regulatory control of pilus expression in *Escherichia coli* by *Lrp* and DNA methylation: model building based on analysis of *pap*. *Molecular Microbiology* 6: 2429–35.

Vetter, R. D. 1991. Symbiosis and the evolution of novel trophic strategies: thiotrophic organisms in hydrothermal vents. In: Margulis & Fester (op. cit.), pp. 219–45.

Vossbrinck, C. R., Maddox, J. R., Friedman, S., Debrunner-Vossbrinck, B. A. & Woese, C. R. 1987. Ribosomal RNA sequence suggests microsporidia are extremely ancient eukaryotes. *Nature* 326: 411–14.

Wächtershäuser, G. 1988. Before enzymes and templates: theory of surface metabolism. *Microbiological Reviews* 52: 452–84.

Wächtershäuser, G. 1990. Evolution of the first metabolic cycles. *Proceedings of the National Academy of Sciences, USA* 87: 200–4.

Wächtershäuser, G. 1992. Groundworks for an evolutionary biochemistry: the iron–sulphur world. *Progress in Biophysics and Molecular Biology* 58: 85–201.

Waddington, C. H. 1956. Genetic assimilation of the bithorax phenotype. *Evolution* 10: 1–13.

Ward, J. 1962. A further investigation of the swimming reaction of *Stomphia coccinea*. *American Zoologist* 2: 567.

Webster, G. 1984. The relations of natural forms. In: M.-W. Ho & P. Saunders (eds), *Beyond Neo-Darwinism*, pp. 193–217. Academic Press, London.

Weiner, A. M. & Maizels, N. 1987. 3′ terminal t-RNA-like structures tag genomic RNA molecules for replication: implications for the origin of protein synthesis. *Proceedings of the National Academy of Sciences, USA* 84: 7383–7.

Weismann, A. 1889. *Essays upon Heredity*. Oxford University Press (Clarendon), London.

Weiss, R. L., Kukora, J. R. & Adams, J. 1975. The relationship between enzyme activity, cell geometry, and fitness in *Saccharomyces cerevisiae*. *Proceedings of the National Academy of Sciences, USA* 72: 794–8.

Whatley, J. M. 1982. Ultrastructure of plastid inheritance: green algae to angiosperms. *Biological Reviews* 57: 525–69.

Whitcomb, R. F. 1981. The biology of spiroplasms. *Annual Review of Entomology* 26: 397–425.

White, H. B. 1976. Coenzymes as fossils of an earlier metabolic stage. *Journal of Molecular Evolution* 7: 101–4.

White, H. B. III 1982. Evolution of coenzymes and the origin of pyridine nucleotides. In: J. Everse, B. Anderdson & K. You (eds), *The Pyridine Nucleotide Coenzymes*, pp. 1–17. Academic Press, New York.

Williams, G. C. 1966. *Adaptation and Natural Selection*. Princeton University Press, Princeton.

Wilson, D. S. 1980. *The Natural Selection of Populations and Communities*. Benjamin Cummings, Menlo Park, California.

Wilson, E. O. 1975. *Sociobiology: The New Synthesis*. Harvard University Press, Cambridge, Mass.

Withers, G. S., Fahrbach, S. E. & Robinson, G. E. 1993. Selective neuroanatomical plasticity and division of labour in the honeybee. *Nature* 364: 238–40.

Woese, C. R. 1965. On the evolution of the genetic code. *Proceedings of the National Academy of Sciences, USA* 77: 1083–6.

Woese, C. R. 1967. *The Genetic Code: the Molecular Basis for Genetic Expression*. Harper & Row, New York.

Woese, C. R. 1979. A proposal concerning the origin of life on the planet Earth. *Journal of Molecular Evolution* 13: 95–101.

Woese, C. R. 1987. Bacterial evolution. *Microbiological Reviews* 51: 221–71.

Woese, C. R. & Fox, G. E. 1977. Phylogenetic structure of the prokaryotic domain: the primary kingdoms. *Proceedings of the National Academy of Sciences, USA* 74: 5088–90.

Wolpert, L. 1969. Positional information and the spatial pattern of cellular differentiation. *Journal of Theoretical Biology* 25: 1–447.

Wolpert, L. 1990. The evolution of development. *Biological Journal of the Linnean Society* 39: 109–24.

Wolpert, L. 1991. *The Triumph of the Embryo*. Oxford University Press, Oxford.

Wong, J. T.-F. 1975. A co-evolution theory of the genetic code. *Proceedings of the National Academy of Sciences, USA* 72: 1909–12.

Wong, J. T.-F. 1980. Role of minimization of chemical distances between amino acids in the evolution of the genetic code. *Proceedings of the National Academy of Sciences, USA* 77: 1083–6.

Wong, J. T.-F. 1988. Evolution of the genetic code. *Microbiological Science* 5: 174–81.

Wood, R. J. 1976. Between-family variation in sex ratio in the Trinidad (T-30) strain of *Aedes aegypti* (L.) indicating differences in sensitivity to the meiotic drive gene M^D. *Genetics* 46: 345–61.

Wright, S. 1934. Molecular and evolutionary theories of dominance. *American Naturalist* 63: 24–53.

Wright, S. 1939. The distribution of self-sterility alleles in populations. *Genetics* 24: 538–52.

Yanagawa, H., Ogawa, Y., Ueno, M., Sasaki, K. & Sato, T. 1990. A novel minimum ribozyme with oxidoreduction activity. *Biochemistry* 29: 10585–9.

Zaug, A. J. & Cech, T. R. 1986. The intervening sequence RNA of *Tetrahymena* is an enzyme. *Science* 231: 470–5.

Zielinski, W. S. & Orgel, L. E. 1987. Autocatalytic synthesis of a tetranucleotide analogue. *Nature* 327: 346–7.

Zull, J. E. & Smith, S. K. 1990. Is genetic code redundancy related to retention of structural information in both DNA strands? *Trends in Biochemical Sciences* 15: 257–61.

Author Index

Subject Index

A

Acraea 173
adenine, abiotic synthesis of 30
Aedes 173
Alchemilla 164, 166
amino acids, prebiotic synthesis of 28–30
anisogamy
 origin of 159–62
 and intragenomic conflict 173–5
Antirrhinum 229, 237
aphids, symbiotic in insects 195
Arabidopsis 227–9, 235, 237
Archaebacteria 122–4
Archaezoa 122–4, 149, 152, 157
Atta 191
autocatalysis 20–1, 31–6, 105
 and CO_2 fixation 106–7
autocatalytic nets 67–72
autotrophy 106–8

B

B chromosomes 175–6
bacteriophage 47, 63, 66, 197
 host-induced modification in 221
Barbulanympha 149
bauplan 243
Bayer's patches 110–13
Blastiocladiella 206
Broca's area 296–7, 300

C

C-value paradox 179–81
Cambrian explosion 203, 205
cell cycle 207–9
cell differentiation 214–15
cell heredity 220–3, 247–50
chemoton, the 20–4, 102, 105
Chlamydomonas 144, 160–1, 210
 inheritance of organelles in 183

chloroplasts 122, 124
 origin of 137–42
 uniparental inheritance of 161–2,
 182–3
Chromista 142
chromosomes
 bacterial 132
 evolution of form in 181–2
 origin of 114–17
code, genetic 82
 modifications of 81–3
 origin of 84–94
 redundancy of 88–9
coenzymes 64–5, 89–92
compartments 53–4, 57–8, 99
complexity 3–6
copia 178
coral reefs 190
creole languages 305–7
Crepis 176
crossing-over 153–6
Cyanophora 196

D

Danaus 173–4
division of labour 12–13, 58, 212, 268–70
DNA, 12, 42
 double-strand damage 152, 153–4
 mitochondria 278
 origin of 117
 replication 109
 single-strand damage 156
 timing of replication 207
dolphins, language in 298–300
Drosophila
 bristle patterns in 233
 gradients in egg 204, 230, 234–5
 homeotic mutants in 250
 inversions in 182
 segment formation in 233–4, 235–7